Applied analy[illegible] equations

Cambridge Texts in Applied Mathematics

Maximum and Minimum Principles
M.J. Sewell

Solitons
P.G. Drazin and R.S. Johnson

The Kinematics of Mixing
J.M. Ottino

Introduction to Numerical Linear Algebra and Optimisation
Philippe G. Ciarlet

Integral Equations
David Porter and David S.G. Stirling

Perturbation Methods
E.J. Hinch

The Thermomechanics of Plasticity and Fracture
Gerard A. Maugin

Boundary Integral and Singularity Methods for Linearized Viscous Flow
C. Pozrikidis

Nonlinear Wave Processes in Acoustics
K. Naugolnykh and S. Ostrovsky

Nonlinear Systems
P.G. Drazin

Stability, Instability and Chaos
Paul Glendinning

Viscous Flow
H. Ockendon and J.R. Ockendon

Applied Analysis of the Navier-Stokes Equations
Charles R. Doering and J.D. Gibbon

Applied analysis of the Navier-Stokes equations

CHARLES R. DOERING

Clarkson University

and

Los Alamos National Laboratory

J. D. GIBBON

Imperial College of Science, Technology and Medicine

Published by the Press Syndicate of the University of Cambridge
The Pitt Building, Trumpington Street, Cambridge CB2 1RP
40 West 20th Street, New York, NY 10011-4211, USA
10 Stamford Road, Oakleigh, Melbourne 3166, Australia

First published 1995

Library of Congress Cataloging-in-Publication Data

Doering Charles R.
Applied analysis of the Navier-Stokes equations / C.R. Doering, J.D. Gibbon.
p. cm. – (Cambridge texts in applied mathematics)
Includes bibliographical references and index.
ISBN 0-521-44557-4. – ISBN 0-521-44568-X (pbk.)
1. Navier-Stokes equations. I. Gibbon, J. D. II. Title.
III. Series.
QA374.D58 1995
532′.0527′01515353–dc20 94-21610
CIP

A catalog record for this book is available from the British Library.

ISBN 0-521-44557-4 Hardback
ISBN 0-521-44568-X Paperback

Cover figure: Final-state vorticity in a simulation of free decay in the two-dimensional Navier–Stokes equations. Reprinted from Matthaeus, Stribling, Martinez, Oughton and Montgomery (1991), 'Decaying, two-dimensional, Navier–Stokes turbulence at very long times', *Physica D*, 51, pp. 531–538, by permission of Elsevier Science Publishing.

Transferred to digital printing 2001

TAG

To
Paula
and
Sheila

Contents

Preface

This book is not meant to be a review or a reference work, nor did we write it as a research monograph. It is not a text on fluid mechanics, and it is not an analysis course book. Rather, our goal is to outline one specific challenge that faces the next generation of applied mathematicians and mathematical physicists. The problem, which we believe is not widely appreciated in these communities, is that it is not at all certain whether one of the fundamental models of classical mechanics, of wide utility in engineering applications, is actually self-consistent.

The suspect model is embodied in the Navier-Stokes equations of incompressible fluid dynamics. These equations are nothing more than a continuum formulation of Newton's laws of motion for material "trying to get out of its own way." They are a set of nonlinear partial differential equations which are thought to describe fluid motions for gases and liquids, from laminar to turbulent flows, on scales ranging from below a millimeter to astronomical lengths. Only for the simplest examples are they exactly soluble, though, usually corresponding to laminar flows. In many important applications, including turbulence, they must be modified and matched, truncated and closed, or otherwise approximated analytically or numerically in order to extract any predictions. On its own this is not a fundamental barrier, for a good approximation can sometimes be of equal or greater utility than a complicated exact result.

The issue is that it has never been shown that the Navier-Stokes equations, in three spatial dimensions, possess smooth solutions starting from arbitrary initial conditions, even very smooth, physically reasonable initial conditions. It is possible that the equations produce solutions which exhibit finite-time singularities. If this occurs, then subsequent evolution may be nonunique, violating the fundamental tenets of Newtonian determinism for this model. Furthermore, finite-time singularities in the solutions signal that the equations are generating structures on arbitrarily small scales, contradicting the separation-of-scales assumption used to derive the hydrodynamic equations from microscopic models. It turns out that the nonlinear terms that can't be controlled mathematically are precisely those describing what is presumed to be the basic physical mechanism for the generation of turbulence, namely vortex stretching. So what may appear to applied scientists to be mathematical formalities, i.e., questions of existence and uniqueness and regularity, are actually intimately tied up with the efficacy of the Navier-Stokes equations as a model for fluid turbulence. Whether or not the equations actually do display these pathologies remains an open problem: It's never been proved one way or the other.

In this book we have tried to lay out the details of this quandary. In the first four chapters we introduce the Navier-Stokes equations together with some fundamental ideas about stability, turbulence, and dynamical systems in general. The remaining chapters deal with the associated mathematical issues, starting from what we can prove about existence and uniqueness and leading to the limits of what is known about their regularity. Our goals are to show how far we can go toward establishing regularity for solutions of the Navier-Stokes equations, to show some encouraging results of rigorous treatments of Navier-Stokes dynamics, and to show the shortcomings and limitations of the analysis. With the deepest respect and admiration for those who identified and initiated investigations into these mathematical issues, we recognize that this topic has remained closed to the mainstream of applied mathematics and mathematical physics, due in large part to the technical nature of the investigations, often phrased in the unfamiliar language of abstract functional analysis. The attempt we have made is to present some of the techniques and results of these studies in a familiar context, explaining and developing the tools as we proceed.

We have intended this book for graduate students and researchers in applied mathematics and theoretical/mathematical physics who want an introductory and detailed presentation of the methods, successes, and limitations of formal analysis of the Navier-Stokes equations. We presume a level of mathematical sophistication approximately at the level of a first year (UK) or second year (USA) mathematics graduate student, or of a similarly prepared physics or engineering student. The greatest part of the book is based on elementary ideas from Newtonian mechanics, real analysis, Fourier analysis, ordinary differential equations, and linear algebra. We have attempted to explain each step of the essential calculations and proofs, with the thought that this is what it takes not to lose or dishearten motivated, curious, and diligent readers.[1] In its entirety, this book would be a suitable text for a one semester graduate level special topics course. A selection of the material, say, Chapters 1-3, 5-6, and 8, comprising the essential elements of the statement and analysis of the problem, the methods and the results, could be covered in a normal British university term or American academic quarter.

Many people have contributed to our understanding of the contents of this book, and in particular we acknowledge fruitful discussions, collaborations, and/or interactions with M. Bartuccelli, C. Foias, D. Holm, S. Malham, and R. Temam. Marieka Fisher efficiently and skillfully assisted with the manuscript preparation. Special thanks go to Mark Alderson, James Robinson, and Edriss Titi who thoroughly and critically read through the manuscript. Finally, and most of all, we thank Peter Constantin and Dave Levermore who, over the last seven years, have patiently, expertly, and cheerfully taught us much of the material in this book, making the complicated simple, and the obscure clear. Much credit goes to them for what is worthwhile about this book. The blame for its flaws rests squarely on us.

[1] Many such readers will notice our preoccupation with the derivation and manipulation of inequalities, rather than equalities. At first this may seem to be taking the easy route to obtaining relatively weak results, but several comments are in order. First, it is often the case in theoretical physics and applied mathematics that "exact" calculations are performed for specific solutions of a problem, or for approximations of a problem. We are often concerned here with results for all (or a class of) solutions, without approximations. This means that bounds or "estimates," rather than exact values, are the only things that are realistically possible to prove. Furthermore, although one might think that it's easier to compute with (imprecise) inequalities than with (precise) equalities, actually it is just the opposite: With equalities there is one right answer, while with inequalities there are an infinite number of right answers. We are thus faced with the additional challenge of producing the best right answer!

1

The equations of motion

1.1 Introduction

The Navier-Stokes equations of fluid dynamics are a formulation of Newton's laws of motion for a continuous distribution of matter in the fluid state, characterized by an inability to support shear stresses. We will restrict our attention to the incompressible Navier-Stokes equations for a single component Newtonian fluid. Although they may be derived systematically from the microscopic description in terms of a Boltzmann equation, albeit with some additional fundamental assumptions, in this chapter we present a heuristic derivation designed to illustrate the elements of the physics contained in the equations.

1.2 Euler's equations for an incompressible fluid

First we consider an ideal inviscid fluid. The dependent variables in the so-called Eulerian description of fluid mechanics are the fluid density $\rho(\mathbf{x}, t)$, the velocity vector field $\mathbf{u}(\mathbf{x}, t)$, and the pressure field $p(\mathbf{x}, t)$. Here $\mathbf{x} \in \mathbf{R}^d$ is the spatial coordinate in a d-dimensional region of space (d typically takes values 2 or 3, with a default value of 3 in this chapter). An infinitesimal element of the fluid of volume δV located at position $\mathbf{x}$ at time t has mass $\delta m = \rho(\mathbf{x}, t)\delta V$ and is moving with velocity $\mathbf{u}(\mathbf{x}, t)$ and momentum $\delta m\, \mathbf{u}(\mathbf{x}, t)$. The normal force directed into the infinitesimal volume across a face of area $\mathbf{n}\,\delta a$ centered at $\mathbf{x}$, where $\mathbf{n}$ is the outward directed unit vector normal to the face, is $-\mathbf{n}p(\mathbf{x}, t)\delta a$. The pressure is the magnitude of the force per unit area, or normal stress, imposed on elements of the fluid from neighboring elements. These definitions are illustrated in Figure 1.1.

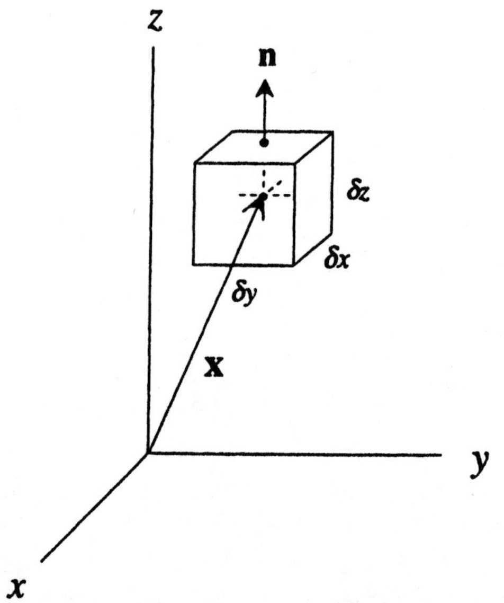

Fig. 1.1. A fluid element of volume $\delta V = \delta x \delta y \delta z$ located at position $\mathbf{x}$. The top surface's outward pointing normal $\hat{\mathbf{n}}$ is shown.

The fundamental kinematic principle is contained in the notion of the convective derivative. On the one hand, the rate of change of a quantity given by the function $f(\mathbf{x}, t)$ at a fixed point $\mathbf{x}$ in space is simply the partial derivative with respect to time:

$$\left(\frac{df(\mathbf{x},t)}{dt}\right)_{\text{fixed position}} = \lim_{\delta t \to 0} \frac{f(\mathbf{x}, t+\delta t) - f(\mathbf{x}, t)}{\delta t} = \frac{\partial f(\mathbf{x}, t)}{\partial t}. \quad (1.2.1)$$

On the other hand, the rate of change of the same quantity at $\mathbf{x}$, as measured by an observer moving with velocity $\mathbf{u}$, is

$$\begin{aligned}\left(\frac{df(\mathbf{x},t)}{dt}\right)_{\text{moving}} &= \lim_{\delta t \to 0} \frac{f(\mathbf{x}+\mathbf{u}\delta t, t+\delta t) - f(\mathbf{x}, t)}{\delta t} \\ &= \frac{\partial f(\mathbf{x}, t)}{\partial t} + \mathbf{u} \cdot \nabla f(\mathbf{x}, t). \end{aligned} \quad (1.2.2)$$

We refer to this rate of change with respect to an observer moving with the fluid, as the *convective derivative* and denote it by d/dt. That is, for a function of both $\mathbf{x}$ and t,

$$\frac{df(\mathbf{x},t)}{dt} := \frac{\partial f(\mathbf{x}, t)}{\partial t} + \mathbf{u} \cdot \nabla f(\mathbf{x}, t). \quad (1.2.3)$$

There is no ambiguity in the definition of the time derivative for functions of time alone, where the standard notation d/dt will be used.

The fundamental equations of motion for a fluid system characterized by $\rho(\mathbf{x},t)$, $\mathbf{u}(\mathbf{x},t) = (u_1,u_2,u_3)$ and $p(\mathbf{x},t)$ come from three different distinct considerations: conservation of mass, Newton's second law, and material properties.

Consider the volume δV of an element of mass δm as the system evolves. Conservation of mass means that δm doesn't change for this element. If the element compresses or expands then the volume and density will change, but the mass is fixed:

$$\frac{d\delta m}{dt} = 0. \tag{1.2.4}$$

The rate of change of the volume occupied by δm is obtained as follows. For a rectangular volume $\delta V = \delta x\delta y\delta z$ we write

$$\frac{d\delta V}{dt} = \frac{d\delta x}{dt}\delta y\delta z + \frac{d\delta y}{dt}\delta x\delta z + \frac{d\delta z}{dt}\delta x\delta y. \tag{1.2.5}$$

The length elements increase or decrease according to the relative velocity of their endpoints. The rate of change of the length δx is

$$\frac{d\delta x}{dt} = u_1(x+\delta x/2,y,z,t) - u_1(x-\delta x/2,y,z,t) = \frac{\partial u_1}{\partial x}\,\delta x, \tag{1.2.6}$$

and likewise for the other components. Combined with equation (1.2.5), this gives

$$\frac{d\delta V}{dt} = (\nabla\cdot\mathbf{u})\,\delta V. \tag{1.2.7}$$

Hence the divergence of the velocity vector field is the local rate of change of the volume of elements of mass. In terms of the density ρ,

$$\frac{d\rho}{dt} = \frac{d}{dt}\frac{\delta m}{\delta V} = -\frac{\delta m}{(\delta V)^2}\frac{d\delta V}{dt} = -\rho\nabla\cdot\mathbf{u}. \tag{1.2.8}$$

Using the definition of the convective derivative, we see that conservation of mass manifests itself as the *continuity equation*

$$0 = \frac{\partial\rho}{\partial t} + \mathbf{u}\cdot\nabla\rho + \rho\nabla\cdot\mathbf{u} = \frac{\partial\rho}{\partial t} + \nabla\cdot(\rho\mathbf{u}). \tag{1.2.9}$$

Newton's second law of motion, which states that the rate of change of momentum equals the net applied force, can be applied to each element

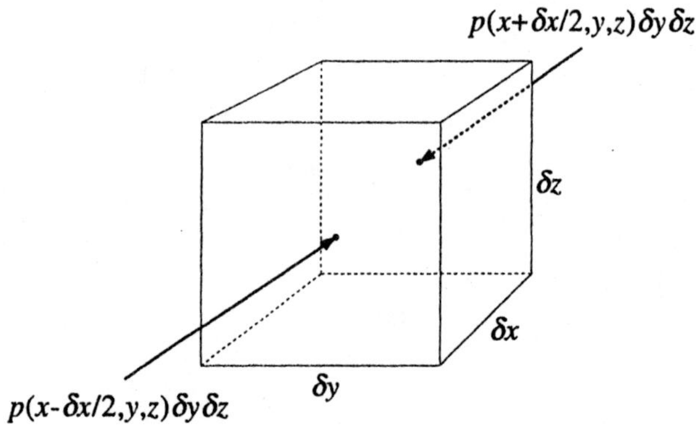

Fig. 1.2. The pressure force acting on the front and rear faces of a fluid element.

of mass in the fluid. In the absence of any externally applied forces, the net force δF acting on each element of mass is due to the pressure field. The component of force in the x-direction, as illustrated in Figure 1.2, is

$$\delta F_1 = p\left(\mathbf{x} - \hat{\mathbf{i}}\delta x/2, t\right)\delta y\delta z - p\left(\mathbf{x} + \hat{\mathbf{i}}\delta x/2, t\right)\delta y\delta z = -\frac{\partial p}{\partial x}\delta V. \quad (1.2.10)$$

Similar expressions hold for the y and z components of the force. Hence Newton's second law for the element of fluid mass δm at position δx is,

$$\frac{d}{dt}(\delta m\, \mathbf{u}(\mathbf{x}, t)) = \delta \mathbf{F} = -\delta V\, \nabla p. \quad (1.2.11)$$

Recalling the equation of conservation of mass (1.2.4) and the definition of the convective derivative (1.2.3) and dividing through by δm we find *Euler's equations*

$$\frac{\partial \mathbf{u}}{\partial t} + \mathbf{u} \cdot \nabla \mathbf{u} = -\frac{1}{\rho}\nabla p. \quad (1.2.12)$$

Combined, the continuity equation (1.2.9) and Euler's equations (1.2.12) provide $d+1$ evolution equations for the $d+2$ dependent variables (ρ, p and the d components of $\mathbf{u}$). What remains is to provide a connection between the density and pressure. Typically this is in the form of a thermodynamic equation of state. For example, in an ideal gas at constant temperature, $p \sim \rho$. If temperature variations are to be accounted for, then the pressure may become a function of both the local density and the local temperature and a further evolution equation for the temperature must be supplied. This matter will be taken up in section 1.5. A significant simplification is achieved by considering fluids which are effectively incompressible. Mathematically the condition of incompressibility is simply

$$\nabla \cdot \mathbf{u} = 0. \tag{1.2.13}$$

Physically, this constraint restricts applicability to problems where all the relevant velocities are much less than the speed of sound in the fluid. The continuity equation (1.2.9) then implies that the convective derivative of the density vanishes, so the density of each fluid element never changes from its initial value. This, in turn, implies that an initially homogeneous (constant density) fluid remains so:

$$\rho(\mathbf{x}, 0) = \text{constant} \qquad \Rightarrow \qquad \rho(\mathbf{x}, t) = \text{constant}. \tag{1.2.14}$$

Euler's equations for an incompressible homogeneous fluid are

$$\frac{\partial \mathbf{u}}{\partial t} + \mathbf{u} \cdot \nabla \mathbf{u} + \frac{1}{\rho} \nabla p = 0 \tag{1.2.15}$$

$$\nabla \cdot \mathbf{u} = 0, \tag{1.2.16}$$

where the density is now a parameter. These are $d+1$ equations for the $d+1$ unknowns (p and the d components of $\mathbf{u}$). The pressure is determined by the velocity vector field, as is seen by taking the divergence of (1.2.15), commuting the space and time derivatives, and using the divergence-free condition on $\mathbf{u}$ from (1.2.16):

$$\Delta p = -\rho \nabla \cdot (\mathbf{u} \cdot \nabla \mathbf{u}) = -\rho\, u_{i,j} u_{j,i}. \tag{1.2.17}$$

The pressure field is a solution of Poisson's equation with a source which is quadratic in the derivatives of the velocities; the pressure is a nonlocal functional of the instantaneous flow configuration. As a result of the incompressibility condition, the pressure is the stress applied by neighboring parts of the fluid on one another in an attempt to "push each other out of the way." These forces are propagated instantaneously (the speed of sound is effectively infinite) and correlated over long ranges (the kernel of the inverse of the Laplacian decays as r^{-1} in three dimensions). Combining equations (1.2.15) and (1.2.17), Euler's equations may be thought of as a nonlocal evolution equation for $\mathbf{u}$ alone.[1] They are not completely posed, however, until appropriate boundary conditions are specified. Boundary conditions are necessary for both $\mathbf{u}$ in (1.2.15) and p in (1.2.17), and are determined by the physics of the problem at hand.

If the fluid is confined to a fixed region of space Ω bounded by a stationary boundary $\partial\Omega$, then the fluid cannot cross these rigid boundaries. This means that the normal component of the velocity vector field satisfies

$$\mathbf{n} \cdot \mathbf{u}|_{\partial\Omega} = 0, \tag{1.2.18}$$

where $\mathbf{n}$ is the local normal to $\partial\Omega$. To derive boundary conditions for Poisson's equation for the pressure we consider the normal component of (1.2.15) on the boundary. Because, formally,

$$\mathbf{n} \cdot \left.\frac{\partial \mathbf{u}}{\partial t}\right|_{\partial\Omega} = 0 \tag{1.2.19}$$

the pressure satisfies the Neumann boundary conditions

$$\mathbf{n} \cdot \nabla p \,|_{\partial\Omega} = -\rho\, \mathbf{n} \cdot [\mathbf{u} \cdot \nabla \mathbf{u}]|_{\partial\Omega}\,. \tag{1.2.20}$$

For stationary flat boundary surfaces, e.g., the walls of a fixed rectangular box, this condition simplifies to

$$\mathbf{n} \cdot \nabla p|_{\partial\Omega} = 0. \qquad \text{(Euler)} \tag{1.2.21}$$

In these cases these boundary conditions are sufficient to determine p uniquely up to an additive constant (and explicitly, if the relevant Green's functions are known explicitly) for a given instantaneous velocity field $\mathbf{u}$.

Another set of boundary conditions, often used in both theoretical and numerical studies, are periodic boundary conditions where the fluid motion is confined to a 2-torus or 3-torus. Mathematically one imposes periodic boundary conditions on p and each component of $\mathbf{u}$, in each of the d directions, with spatial periods $L_1, ..., L_d$. The technical advantage of periodic boundary conditions is that they allow for the study of a

[1] This is the view taken in the mathematical literature. Together, (1.2.15), (1.2.16), and (1.2.17) may be written in the compact form

$$\mathbf{P}\left\{\frac{d\mathbf{u}}{dt}\right\} = 0,$$

where the operator $\mathbf{P}$ is a projection onto divergence-free vector fields. That is, for a vector field $\mathbf{v}$ the formal expression for the projection is

$$\mathbf{P}\{\mathbf{v}\} = \mathbf{v} - \nabla\Delta^{-1}(\nabla \cdot \mathbf{v}),$$

where Δ^{-1} is the inverse Laplacian with some appropriate boundary conditions. In this notation the pressure is explicitly recognized as an artifact of the incompressibility condition.

finite volume of fluid both with translation invariance and without either the physical or mathematical complications of rigid boundaries. The implications of these complications will become apparent in later chapters. For either choice of boundary conditions, the mathematical challenge is to show that along with appropriate initial conditions (say, the flow configuration $\mathbf{u}(\mathbf{x}, 0)$ at $t = 0$), Euler's equations are an apparently well posed problem for the time evolution of the velocity vector field.

1.3 Energy, body forces, vorticity, and enstrophy

Consider an ideal incompressible fluid in the volume Ω bounded by $\partial\Omega$, with either rigid or periodic boundary conditions. The kinetic energy in the fluid is

$$\int_\Omega \left(\frac{1}{2}\rho\,|\mathbf{u}|^2\right) d^dx = \frac{1}{2}\rho\,\|\mathbf{u}\|_2^2, \tag{1.3.1}$$

where we have introduced the notation $\|\cdot\|_2$ for the norm in $L^2(\Omega)$, the Hilbert space of square integrable functions. The usual law of conservation of energy applies to an ideal incompressible fluid, and it can be derived from Euler's equations. Differentiating $\frac{1}{2}\rho\,\|\mathbf{u}\|_2^2$ with respect to time and using the equation of motion (1.2.15), we find

$$\begin{aligned} \frac{d}{dt}\left(\frac{1}{2}\rho\,\|\mathbf{u}\|_2^2\right) &= \rho \int_\Omega \mathbf{u}\cdot\frac{\partial \mathbf{u}}{\partial t}\, d^dx \\ &= -\rho \int_\Omega \mathbf{u}\cdot\left[\mathbf{u}\cdot\nabla\mathbf{u} + \frac{1}{\rho}\nabla p\right] d^dx. \end{aligned} \tag{1.3.2}$$

Noting the divergence-free condition on $\mathbf{u}$,

$$\mathbf{u}\cdot(\mathbf{u}\cdot\nabla\mathbf{u}) = \frac{1}{2}\mathbf{u}\cdot\nabla|\mathbf{u}|^2 = \frac{1}{2}\nabla\cdot\left(\mathbf{u}|\mathbf{u}|^2\right) \tag{1.3.3}$$

and

$$\mathbf{u}\cdot\nabla p = \nabla\cdot(\mathbf{u}p). \tag{1.3.4}$$

The divergence theorem applied to the last line of (1.3.2) yields

$$\frac{d}{dt}\left(\frac{1}{2}\rho\,\|\mathbf{u}\|_2^2\right) = -\int_{\partial\Omega}\left(\frac{1}{2}\rho\,|\mathbf{u}|^2 + p\right)\mathbf{u}\cdot\mathbf{n}\,da. \tag{1.3.5}$$

For stationary rigid boundary conditions $\mathbf{u}\cdot\mathbf{n}$ vanishes on $\partial\Omega$ so the surface integral vanishes. And for periodic boundary conditions, any such "surface" integral vanishes identically (because there really is no surface). Hence in either case,

$$\frac{d}{dt}\left(\frac{1}{2}\rho\,\|\mathbf{u}\|_2^2\right)=0 \tag{1.3.6}$$

and the total kinetic energy is conserved. This is natural because we have not taken any dissipative effects into account, and there is no external work being done on the system. Mathematically, we say that the L^2 norm is conserved.

If an external force is applied to the fluid, then Euler's equations for an incompressible fluid become

$$\frac{\partial \mathbf{u}}{\partial t}+\mathbf{u}\cdot\nabla\mathbf{u}+\frac{1}{\rho}\nabla p=\frac{1}{\rho}\,\mathbf{f}(\mathbf{x},t) \tag{1.3.7}$$

$$\nabla\cdot\mathbf{u}=0, \tag{1.3.8}$$

where the "body force" $\mathbf{f}(\mathbf{x},t)$ is the applied force per unit volume. The kinetic energy then evolves according to

$$\frac{d}{dt}\left(\frac{1}{2}\rho\,\|\mathbf{u}\|_2^2\right)=\int_\Omega \mathbf{u}\cdot\mathbf{f}\,d^dx. \tag{1.3.9}$$

The source or sink of energy in (1.3.9) is the power expended or absorbed by the body force.

If the force field is the gradient of a potential per unit volume, i.e., if

$$\mathbf{f}=-\nabla\phi, \tag{1.3.10}$$

then the net work done by the body force vanishes identically for either stationary rigid or periodic boundary conditions:

$$\int_\Omega \mathbf{u}\cdot(-\nabla\phi)\,d^dx=-\int_{\partial\Omega}\phi\mathbf{u}\cdot\mathbf{n}\,da. \tag{1.3.11}$$

In the case of a gradient body force, the body force can be absorbed into the pressure term. The evolution equation (1.3.7) is simply

$$\frac{\partial \mathbf{u}}{\partial t}+\mathbf{u}\cdot\nabla\mathbf{u}+\frac{1}{\rho}\nabla(p+\phi)=0, \tag{1.3.12}$$

and the solution for $\mathbf{u}(\mathbf{x},t)$ is the same whether or not ϕ is present; the potential just renormalizes the pressure field. Hence when we include external body forces, we will typically consider nongradient fields, i.e., those with a nonvanishing curl in a simply connected domain.

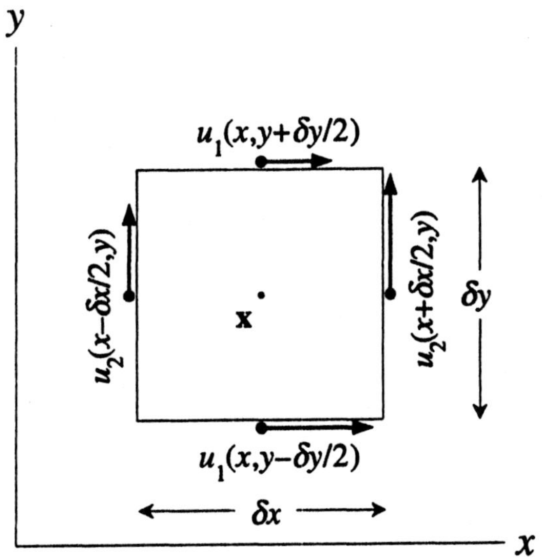

Fig. 1.3. Components of the velocity on the faces of a fluid element contributing to the average angular velocity in the z-direction.

Besides the local velocity of the fluid, another important kinematic quantity is the local angular velocity. The standard measure of this angular velocity is the vorticity $\boldsymbol{\omega}$, defined by

$$\boldsymbol{\omega} = \nabla \times \mathbf{u}. \tag{1.3.13}$$

To see how this is related to the angular velocity of a piece of the fluid, consider a small rectangular mass δm located at $\mathbf{x}$ as in Figure 1.3. Then the average z-component, $\langle \Omega_3 \rangle$, of the angular velocity of the particle's edges about the point $\mathbf{x}$ is

$$\begin{aligned} \langle \Omega_3 \rangle &= \frac{1}{4}\left[\frac{u_2\left(x+\delta x/2, y\right)}{\delta x/2} - \frac{u_1\left(x, y+\delta y/2\right)}{\delta y/2} \right. \\ &\quad \left. - \frac{u_2(x-\delta x/2, y)}{\delta x/2} + \frac{u_1(x, y-\delta y/2)}{\delta y/2}\right] \\ &= \frac{1}{2}\left[\frac{\partial u_2}{\partial x} - \frac{\partial u_1}{\partial y}\right] \\ &= \frac{1}{2}\omega_3. \end{aligned} \tag{1.3.14}$$

An evolution equation for the vorticity is derived by taking the curl of

Euler's equation for the velocity vector field. Using standard $3d$ vector calculus identities along with the divergence free condition on $\mathbf{u}$ we obtain, from (1.2.15) and (1.2.16),

$$\frac{\partial \boldsymbol{\omega}}{\partial t} + \mathbf{u} \cdot \nabla \boldsymbol{\omega} = \boldsymbol{\omega} \cdot \nabla \mathbf{u}. \qquad (1.3.15)$$

(If a body force $\mathbf{f}$ is present, then there is also a $\rho^{-1} \nabla \times \mathbf{f}$ on the right-hand side above.) The vorticity evolution equation contains some very important physics which plays a fundamental role in understanding the challenges of fluid turbulence and its mathematical manifestations. The complex fluid motions associated with turbulent flows are often described in terms of "eddies," which are the vortices – local concentrations of vorticity – in a fluid. The vorticity equation also provides the first indication of the fundamental difference between fluid flows in two and three spatial dimensions.

First consider the $2d$ problem. If the velocity vector field is confined to the $x - y$ plane, where $\mathbf{u}(\mathbf{x}, t) = \hat{\mathbf{i}} u_1(x, y, t) + \hat{\mathbf{j}} u_2(x, y, t)$, then the curl only has a z component ω_3, which we will unambiguously refer to as the scalar ω. Only the z component of (1.3.15) does not vanish identically, although every component of the $\boldsymbol{\omega} \cdot \nabla \mathbf{u}$ term does vanish identically. In $2d$ the scalar vorticity ω evolves according to

$$\frac{\partial \omega}{\partial t} + \mathbf{u} \cdot \nabla \omega = 0, \qquad (1.3.16)$$

which is more transparently written in terms of the convective derivative,

$$\frac{d\omega}{dt} = 0. \qquad (1.3.17)$$

The vorticity of each particle of the fluid is a constant of the motion. There is no internal mechanism in the $2d$ Euler's equations for angular velocity to be transferred between different parts of the fluid. Individual parts of the fluid are transported by the flow field $\mathbf{u}$, but the local angular velocity associated with each part remains the same. Initially distinct vortices (localized concentrations of vorticity) are simply transported in the resulting flow field, interacting in possibly complicated ways, but maintaining their identity both in form and in magnitude.

The full $3d$ problem is very different. In terms of the convective derivative, the vorticity equation (1.3.15) is

$$\frac{d\boldsymbol{\omega}}{dt} = \boldsymbol{\omega} \cdot \nabla \mathbf{u}. \qquad (1.3.18)$$

The $\boldsymbol{\omega} \cdot \nabla \mathbf{u}$ term gives rise to a phenomenon referred to as *vortex stretching*.

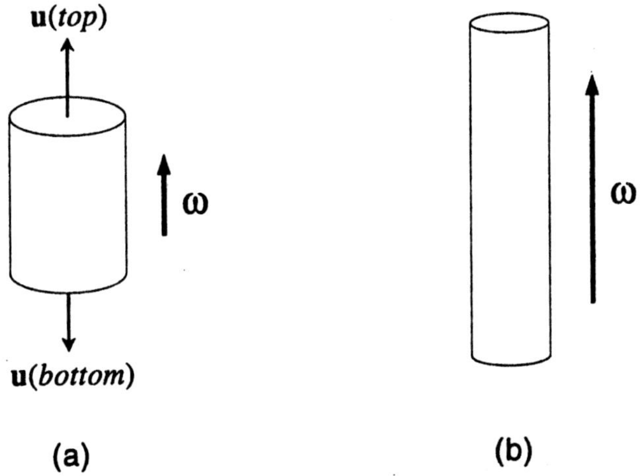

Fig. 1.4. The vortex stretching mechanism. When $\boldsymbol{\omega}\cdot\nabla\mathbf{u}$ has a component parallel to $\boldsymbol{\omega}$, as in (*a*), the fluid element is stretched in the direction of the vorticity. The resulting decrease in the element's moment of inertia, illustrated in (*b*), leads to an increase in the amplitude of the vorticity.

Vortex stretching acts as an amplifier of vorticity for individual particles of the fluid. This $3d$ mechanism is just elementary conservation of angular momentum. To see how the vortex stretching term operates, consider a cylindrical shaped piece of fluid with angular momentum along the z-axis as shown in Figure 1.4a. If the flow field is momentarily configured so that the u_3 is locally increasing in the z-direction, then the cylinder will tend to be stretched as shown in Figure 1.4b. Conservation of angular momentum for the mass in the cylinder then implies an increase in the magnitude of the angular velocity. This can work the other way, too. The magnitude of the vorticity will decrease if the flow conspires to increase the fluid particle's relevant moment of inertia. The vortex stretching mechanism does not increase or decrease the total angular velocity: That is, global conservation of angular momentum leads to conservation of the integrated vorticity, but it can lead to local intensification of the vorticity amplitude.

A global measure of the vorticity content of a flow is the square of its L^2 norm, called[1] the *enstrophy*:

$$\|\boldsymbol{\omega}\|_2^2 = \int_\Omega |\boldsymbol{\omega}|^2 \, d^d x. \tag{1.3.19}$$

[1] The word "enstrophy," coined by George Nickel, is from the Greek noun "strophe" meaning "a turn."

For periodic or stationary rigid boundary conditions in $2d$, the enstrophy is conserved by the Euler equations. In fact, in $2d$ all the moments of ω are conserved with these boundary conditions:

$$\begin{aligned}
\frac{d}{dt}\int_\Omega \omega^n \, d^2x &= n\int_\Omega \omega^{n-1}\frac{\partial\omega}{\partial t}\, d^2x \\
&= -n\int_\Omega \omega^{n-1}\mathbf{u}\cdot\nabla\omega \, d^2x \\
&= -\int_\Omega \nabla\cdot(\omega^n\mathbf{u})\, d^2x \\
&= -\int_{\partial\Omega} \mathbf{n}\cdot(\omega^n\mathbf{u})\, dl \\
&= 0.
\end{aligned} \tag{1.3.20}$$

In $3d$, (1.3.20) is not valid. Enstrophy may be created or destroyed by the vortex stretching term $\boldsymbol{\omega}\cdot\nabla\mathbf{u}$.

•

1.4 Viscosity, the stress tensor, and the Navier-Stokes equations

Viscosity is the tendency of a fluid to resist shearing motions. As such, it is a frictional force with its origins in the microscopic interactions between the atoms or molecules making up the fluid. Its net effect is to dissipate organized, macroscopic forms of energy – the kinetic energy in the flow field – and convert it to the disorganized, microscopic form of energy – heat. Shearing forces in continuum mechanical systems are described by the stress tensor. The tensorial nature of these forces results from the fact that there are two directions associated with each such force, the direction of the force itself and the orientation of the area across which the force acts.

Consider a rectangular shaped portion of fluid, centered on the point (x, y, z) with side lengths $(\delta x, \delta y, \delta z)$, as in Figure 1.5. The component S_{ij} of the stress tensor $\mathbf{S}$ is the force per unit area in the jth direction acting across an area element whose normal is in the ith direction. Forces in the direction of the normal to an area element are associated with the pressure, while those that act in the plane of the element are associated with shear stresses. We adopt the convention that the force acts on the "minus" side (rear, left, bottom) of the area, due to the matter on the "plus" side (front, right, top). Newton's third law implies that forces of equal magnitude and opposite direction act on the "plus" sides due to the matter on the "minus" sides. Adding these forces, the net force on

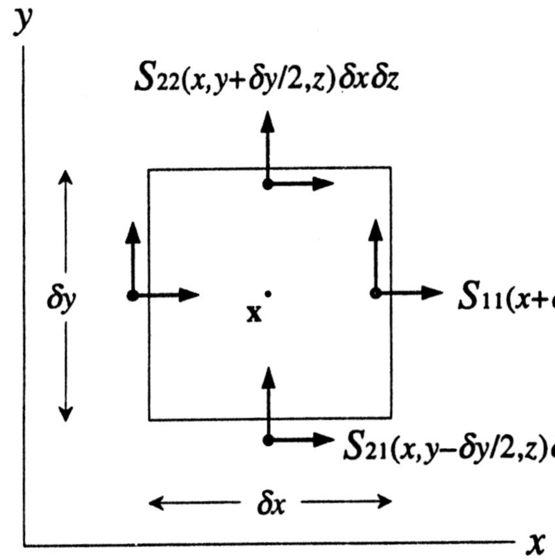

Fig. 1.5. Several components of the stress tensor acting on a fluid element located at $\mathbf{x}$. The forces act on the sides of the faces of the element as indicated by the positions of the vectors. For example, the force acting on the element due to the stress at the bottom face is $-S_{21}(x, y - \delta y/2, z)\delta x \delta z$.

the fluid element in the jth direction is

$$\begin{aligned} \delta F_j &= S_{1j}\left(x + \delta x/2, y, z\right) \delta y \delta z - S_{1j}\left(x - \delta x/2, y, z\right) \delta y \delta z \\ &\quad + S_{2j}\left(x, y + \delta y/2, z\right) \delta x \delta z - S_{2j}\left(x, y - \delta y/2, z\right) \delta x \delta z \\ &\quad + S_{3j}\left(x, y, z + \delta z/2\right) \delta x \delta y - S_{3j}\left(x, y, z - \delta z/2\right) \delta x \delta y. \quad (1.4.1) \end{aligned}$$

Expanding each element of the stress tensor about the center of the element and keeping only the leading terms gives

$$\delta F_j = \left(\frac{\partial S_{1j}(x, y, z)}{\partial x} + \frac{\partial S_{2j}(x, y, z)}{\partial y} + \frac{\partial S_{3j}(x, y, z)}{\partial z} \right) \delta x \delta y \delta z. \quad (1.4.2)$$

Hence the force per unit volume acting at a point in the fluid due to stresses within the fluid is the divergence of the stress tensor, $\delta \mathbf{F}/\delta V = \nabla \cdot \mathbf{S}$.

The stress tensor is always symmetric. The real justification for this assertion is from the definition of the stress tensor in terms of microscopic quantities, but at the level of description adopted in this chapter we can show that the stress tensor must be symmetric in order that local angular accelerations remain finite. To leading order, the z component of torque acting on the volume element due to the stress tensor $\mathbf{S}$ is

$$\begin{aligned} N_3 &= \hat{\mathbf{k}} \cdot \sum_{faces} \mathbf{r} \times \delta \mathbf{F} \\ &= (S_{12} - S_{21})\, \delta x \delta y \delta z \end{aligned} \tag{1.4.3}$$

and the $z - z$ component of the inertia tensor is

$$\begin{aligned} I_{33} &= \int dx \int dy \int \rho \left(x^2 + y^2\right) dz \\ &= \frac{1}{24} \left(\delta x^2 + \delta y^2\right) \rho\, \delta x \delta y \delta z. \end{aligned} \tag{1.4.4}$$

Typical associated angular accelerations are then

$$\frac{N_3}{I_{33}} \sim \frac{S_{12} - S_{21}}{\rho} \frac{1}{\delta x^2 + \delta y^2}. \tag{1.4.5}$$

The necessity of a symmetric stress tensor is then apparent in order to realize a consistent continuum limit as $\delta x \to 0, \delta y \to 0$ and $\delta z \to 0$.

The stress tensor can be decomposed into portions due to the pressure p and the symmetric shear stress tensor T_{ij},

$$S_{ij} = -\delta_{ij}\, p + T_{ij}. \tag{1.4.6}$$

The most general form of the equation of motion for the velocity vector field $\mathbf{u}$, referring to equations (1.2.10), (1.2.11), and (1.2.12), is then

$$\frac{\partial \mathbf{u}}{\partial t} + \mathbf{u} \cdot \nabla \mathbf{u} + \frac{1}{\rho} \nabla p = \frac{1}{\rho} \nabla \cdot \mathbf{T}. \tag{1.4.7}$$

A Newtonian fluid is defined as one in which the shear stress tensor is a linear function of the rate of strain tensor, i.e., the deviation of the fluid motion from a rigid body motion. The rate of strain tensor may be defined as that controlling the evolution of the relative positions of points in a fluid element. Let $\delta \mathbf{x}$ denote infinitesimal displacement of two points in the fluid, one at $\mathbf{x}$ and the other at $\mathbf{x} + \delta \mathbf{x}$. The rate of change of $|\delta \mathbf{x}|^2$ is

$$\begin{aligned} \frac{d}{dt} |\delta \mathbf{x}|^2 &= 2 \delta \mathbf{x} \cdot [\mathbf{u}(\mathbf{x} + \delta \mathbf{x}) - \mathbf{u}(\mathbf{x})] \\ &= \delta x_i \left(\frac{\partial u_i}{\partial x_j} + \frac{\partial u_j}{\partial x_i} \right) \delta x_j \\ &= \delta \mathbf{x} \cdot \mathbf{R} \cdot \delta \mathbf{x}, \end{aligned} \tag{1.4.8}$$

where we have defined the symmetric rate of strain tensor as

$$R_{ij} = \frac{\partial u_i}{\partial x_j} + \frac{\partial u_j}{\partial x_i}. \tag{1.4.9}$$

(The related antisymmetric tensor constructed from the derivatives of the velocity components is equivalent to the vorticity, describing rigid rotational modes of motion that do not involve relative displacements.)

The most general linear isotropic (direction independent) relationship between the symmetric shear stress tensor **T** and the symmetric rate of strain tensor **R** is

$$\mathbf{T} = \alpha \mathbf{R} + \beta \, Tr(\mathbf{R})\mathbf{I}, \tag{1.4.10}$$

where **I** is the unit tensor and the constants α and β are material parameters. The components of the viscous force per unit volume are then

$$(\nabla \cdot \mathbf{T})_i = \alpha \Delta u_i + (2\beta + \alpha)\frac{\partial}{\partial x_i}\nabla \cdot \mathbf{u}. \tag{1.4.11}$$

Together with the condition of incompressibility, $\nabla \cdot \mathbf{u} = 0$, (1.4.7) then becomes the *incompressible Navier-Stokes equations*

$$\frac{\partial \mathbf{u}}{\partial t} + \mathbf{u} \cdot \nabla \mathbf{u} + \frac{1}{\rho}\nabla p = \nu \Delta \mathbf{u}, \tag{1.4.12}$$

$$\nabla \cdot \mathbf{u} = 0. \tag{1.4.13}$$

The material parameter ν is the kinematic viscosity. As with the incompressible Euler equations [recall (1.2.15) and (1.2.16)] the Navier-Stokes equations are four coupled nonlinear partial differential equations for four unknown functions: the three components of **u** and the pressure p. Compared to the incompressible Euler equations, the net effect of the linear coupling between stress and rate of strain is to introduce the "diffusion" term in (1.4.12). The diffusion of momentum between neighboring elements of the fluid is indeed a chief new ingredient in the incompressible Navier-Stokes equations, but there is also the matter of boundary conditions. For whereas the Euler equations are of a hyperbolic form, the addition of viscosity changes the velocity evolution equation into a parabolic form. And taking the divergence of (1.4.12) and using (1.4.13), the pressure is still the solution of the elliptic Poisson equation in (1.2.17).

If the fluid is confined to a fixed region of space Ω bounded by $\partial\Omega$, then the fluid cannot cross the rigid boundaries. This means not only that the normal component of the velocity vector field must vanish on the boundary, as it does in the boundary condition for the Euler equations in (1.2.18), but also that the tangential components of the fluid's velocity are controlled. Physically, the picture is that the microscopic interactions between the fluid particles and the wall are at least as strong as those between fluid particles themselves so that the velocity vector field should be continuous at the wall. These "no-slip" boundary conditions for rigid boundaries are that the fluid at $\partial\Omega$ must move with the prescribed boundary motion. If the velocity of the boundary is given by $\mathbf{U}(\mathbf{x},t)$ for $\mathbf{x} \in \partial\Omega$, then the appropriate boundary conditions for $\mathbf{u}$ are

$$\mathbf{u}|_{\partial\Omega} = \mathbf{U}. \tag{1.4.14}$$

To derive boundary conditions for Poisson's equation for the pressure, we see from the normal component of (1.4.12) on the boundary that for flat stationary ($\mathbf{u} = 0$) boundaries, the pressure satisfies

$$\mathbf{n} \cdot \nabla p|_{\partial\Omega} = \rho\,[\nu \mathbf{n} \cdot \Delta \mathbf{u}]_{\partial\Omega} \qquad \text{(Navier-Stokes)}. \tag{1.4.15}$$

For periodic boundary conditions we impose periodic boundary conditions on p and each component of $\mathbf{u}$.

If an external body force (per unit volume) $\mathbf{f}$ is imposed on the fluid, then the incompressible Navier-Stokes equations become

$$\frac{\partial \mathbf{u}}{\partial t} + \mathbf{u} \cdot \nabla \mathbf{u} + \frac{1}{\rho} \nabla p = \nu \Delta \mathbf{u} + \frac{1}{\rho} \mathbf{f}, \tag{1.4.16}$$

$$\nabla \cdot \mathbf{u} = 0. \tag{1.4.17}$$

With either stationary no-slip or periodic boundary conditions, the kinetic energy's evolution equation follows from (1.4.16) by taking the dot product with $\mathbf{u}$, integrating over Ω, and integrating the $\Delta \mathbf{u}$ term by parts using the boundary conditions:

$$\frac{d}{dt}\left(\frac{1}{2}\rho \|\mathbf{u}\|_2^2\right) = -\nu\rho \|\nabla \mathbf{u}\|_2^2 + \int_\Omega \mathbf{u} \cdot \mathbf{f}\, d^d x, \tag{1.4.18}$$

where

$$\|\nabla \mathbf{u}\|_2^2 = \sum_{i,j=1}^{d} \int_\Omega \left(\frac{\partial u_i}{\partial x_j}\right)^2 d^d x. \tag{1.4.19}$$

The kinetic energy in the flow field changes according to two effects: Energy is dissipated by viscosity (note that the viscous dissipation rate term, $-\nu\rho\|\nabla \mathbf{u}\|_2^2$, is manifestly negative) and power is expended by the external force, which can be of either sign. It is interesting to note that for an incompressible fluid with no-slip or periodic boundary conditions, the viscous energy dissipation rate is proportional to the enstrophy. Indeed, using the definition (1.3.13) of the vorticity we find[1]

$$\begin{aligned}
\|\boldsymbol{\omega}\|_2^2 &= \int_\Omega \varepsilon_{ijk}\varepsilon_{imn}\frac{\partial u_k}{\partial x_j}\frac{\partial u_n}{\partial x_m}\, d^d x \\
&= \int_\Omega (\delta_{jm}\delta_{kn} - \delta_{jn}\delta_{km})\frac{\partial u_k}{\partial x_j}\frac{\partial u_n}{\partial x_m}\, d^d x \\
&= \int_\Omega \left(\frac{\partial u_k}{\partial x_j}\frac{\partial u_k}{\partial x_j} - \frac{\partial u_k}{\partial x_j}\frac{\partial u_j}{\partial x_k}\right) d^d x \\
&= \|\nabla \mathbf{u}\|_2^2 - \int_\Omega \nabla\cdot(\mathbf{u}\cdot\nabla\mathbf{u})\, d^d x, \qquad (1.4.20)
\end{aligned}$$

where the divergence-free condition has been used in the last step for the second term. The integral of the divergence above vanishes for either periodic or no-slip boundaries, and we conclude in those situations that $\|\boldsymbol{\omega}\|_2^2 = \|\nabla\mathbf{u}\|_2^2$. Then the energy equation (1.4.18) becomes

$$\frac{d}{dt}\left(\frac{1}{2}\rho\|\mathbf{u}\|_2^2\right) = -\nu\rho\|\boldsymbol{\omega}\|_2^2 + \int_\Omega \mathbf{u}\cdot\mathbf{f}\, d^d x, \qquad (1.4.21)$$

highlighting the important role of the mean square vorticity: The rate of energy dissipation by viscosity is directly proportional to the enstrophy in the absence of moving boundaries.

The vorticity evolution equation for an incompressible Newtonian fluid is obtained, as before, from the curl of the velocity evolution equation (1.4.16) using standard identities:

$$\frac{\partial\boldsymbol{\omega}}{\partial t} + \mathbf{u}\cdot\nabla\boldsymbol{\omega} = \nu\Delta\boldsymbol{\omega} + \boldsymbol{\omega}\cdot\nabla\mathbf{u} + \frac{1}{\rho}\nabla\times\mathbf{f}. \qquad (1.4.22)$$

The viscosity is a vorticity diffusion coefficient.

In 2*d*, where the vorticity is a scalar and the vortex stretching term vanishes identically, the scalar vorticity ω satisfies

$$\frac{d\omega}{dt} = \nu\Delta\omega + \frac{1}{\rho}\left(\frac{\partial f_2}{\partial x} - \frac{\partial f_1}{\partial y}\right). \qquad (1.4.23)$$

The dynamics of the vorticity consists of convection by the local velocity, diffusion between neighboring fluid elements, and sources or sinks resulting from the body force. Viscosity diminishes the local intensity of vorticity, dissipating "turbulent eddies" if any are present.[2] In 3*d* the

[1] Here we use the Einstein summation convention, summing over repeated indices. The symbol ε_{ijk} is the completely antisymmetric tensor with values $0, \pm 1$ according to whether the indices i, j, k are a cyclic permutation of 1, 2, 3 (+1), an anticyclic permutation (−1), or if any index is repeated (0).

[2] In the presence of rigid walls, viscosity can be a source of vorticity as eddies may be generated and "shed" from no-slip boundaries.

full evolution in (1.4.22) applies and the viscous dissipation of vorticity competes with the production of vorticity by the external body force and amplification by the vortex stretching mechanism.

1.5 Thermal convection and the Boussinesq equations

Although for most of this book we will be concerned with the incompressible Navier-Stokes equations for the velocity field alone, neglecting temperature and density variations, there is one particular generalization that will be considered in some detail. This is the problem of thermal convection (heat transfer) by an incompressible Newtonian fluid. In the first approximation, the local temperature of the fluid may be considered a passive scalar, i.e., a quantity characteristic of each particular fluid element whose space-time evolution is thus controlled by the fluid's motion. Thermal conduction between neighboring fluid elements is then taken into account by including a diffusive term and introducing another material parameter, the thermal diffusion coefficient. The influence of the temperature field on the incompressible fluid's motion is taken into account by introducing a buoyancy force into the velocity evolution equation. The origin of the buoyancy force is the observation that temperature variations typically lead to density variations which, in the presence of a gravitational field, lead to pressure gradients. The inclusion of density variations in the buoyancy force – while neglecting them in the continuity equation – and the neglect of the local heat source due to viscous dissipation constitute the approximate formulation known as the *Boussinesq equations.*

The Boussinesq equations for the velocity vector field, the pressure field and the local temperature $T(\mathbf{x}, t)$ of the fluid are

$$\frac{\partial \mathbf{u}}{\partial t} + \mathbf{u} \cdot \nabla \mathbf{u} + \frac{1}{\rho_0} \nabla p = \nu \Delta \mathbf{u} + \frac{1}{\rho_0} \hat{\mathbf{k}} g \alpha (T - T_0) \tag{1.5.1}$$

$$\nabla \cdot \mathbf{u} = 0 \tag{1.5.2}$$

$$\frac{\partial T}{\partial t} + \mathbf{u} \cdot \nabla T = \kappa \Delta T. \tag{1.5.3}$$

We have introduced the acceleration of gravity $-\hat{\mathbf{k}} g$ ($\hat{\mathbf{k}}$ is the unit vector in the vertical direction taken here as the z-direction), the reference density and temperature ρ_0 and T_0, the thermal expansion coefficient α, and the thermal diffusion coefficient κ. The vertical buoyancy force results from changes in density associated with temperature variations,

$$\rho - \rho_0 = -\alpha(T - T_0). \tag{1.5.4}$$

Density variations are neglected elsewhere in the velocity evolution and continuity equations. Equation (1.5.3) is just a convection-diffusion equation for the temperature field.

The explicit appearance of the reference temperature may be eliminated. On the one hand, in (1.5.1) and (1.5.3) the relative temperature variable $\theta = T - T_0$ can simultaneously replace both T and T_0. On the other hand, the temperature offset in both (1.5.1) and (1.5.3) is really completely arbitrary. Because (1.5.3) only involves derivatives of T, additive constants are irrelevant, and an additive constant to the body force, $\hat{\mathbf{k}}f$, can be absorbed into a redefinition of the pressure, $p \to p - fz$.

Appropriate boundary conditions for the Boussinesq equations depend on the problem at hand. For convection in a horizontal layer between plates held at fixed temperatures, as illustrated in Figure 1.6, the temperature boundary conditions are

$$T|_{z=0} = T_{bottom} \quad \text{and} \quad T|_{z=h} = T_{top} \quad \text{(fixed temperature)}. \tag{1.5.5}$$

Insulating boundary conditions apply to problems where no heat can flow across the boundaries (see below for more about heat flow). At an insulated surface $\partial\Omega$ with local normal unit vector $\mathbf{n}$, the boundary conditions are

$$\mathbf{n} \cdot \nabla T|_{\partial\Omega} = 0 \qquad \text{(insulating)}. \tag{1.5.6}$$

If the top and bottom plates are rigid, no-slip boundaries, then the velocity vector field boundary conditions are

$$\mathbf{u}|_{z=0} = 0 = \mathbf{u}|_{z=h} \qquad \text{(no-slip)}. \tag{1.5.7}$$

For some applications it is appropriate to consider boundaries in the vertical direction which, although letting no matter pass, apply no shear stress to the fluid. For such stress free boundaries the vertical component of the velocity satisfies the Dirichlet boundary conditions

$$u_3|_{z=0} = 0 = u_3|_{z=h} \qquad \text{(stress free)}, \tag{1.5.8}$$

whereas the tangential components satisfy the Neumann conditions

$$\frac{\partial u_1}{\partial z}|_{z=0} = 0 = \frac{\partial u_1}{\partial z}|_{z=h}, \quad \frac{\partial u_2}{\partial z}|_{z=0} = 0 = \frac{\partial u_2}{\partial z}|_{z=h} \quad \text{(stress free)}. \tag{1.5.9}$$

In both these cases, no-slip and stress-free, the boundary conditions in the horizontal directions are determined by the details of the specific

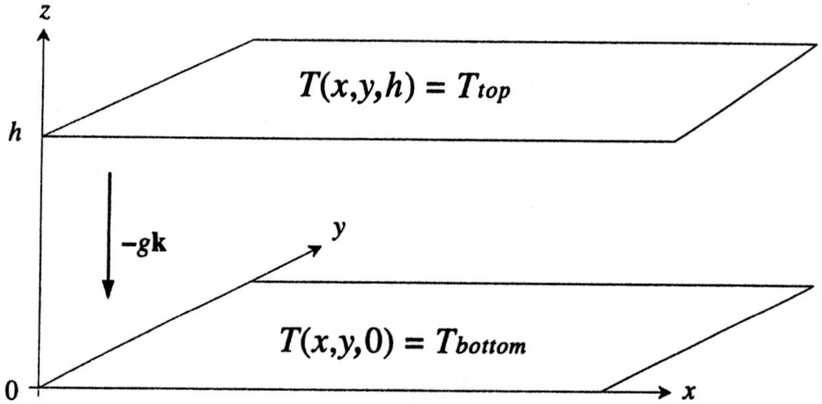

Fig. 1.6. Set-up for the problem of convection between horizontal plates separated by a gap h.

problem at hand. For technical convenience they are often taken as periodic in the x- and y-directions.

As we noted earlier, in analytical studies it may be desirable to avoid altogether the technical difficulties introduced by the presence of rigid boundaries. Another model of convection in a layer which circumvents this problem is to impose a uniform temperature gradient, and then apply periodic conditions in the vertical direction. For a constant vertical temperature gradient $-\delta T/\delta h$, define the temperature variation $\theta(\mathbf{x}, t)$ by

$$T(\mathbf{x}, t) = -\frac{\delta T}{\delta h} z + \theta(\mathbf{x}, t). \tag{1.5.10}$$

The Boussinesq equations become

$$\frac{\partial \mathbf{u}}{\partial t} + \mathbf{u} \cdot \nabla \mathbf{u} + \frac{1}{\rho_0} \nabla p = \nu \Delta \mathbf{u} + \frac{1}{\rho_0} \hat{\mathbf{k}} g \alpha \theta \tag{1.5.11}$$

$$\nabla \cdot \mathbf{u} = 0 \tag{1.5.12}$$

$$\frac{\partial \theta}{\partial t} + \mathbf{u} \cdot \nabla \theta = \kappa \Delta \theta + \frac{\delta T}{\delta h} u_3. \tag{1.5.13}$$

In the above, an additional term from the body force has been absorbed into the pressure,

$$p \to p + \frac{1}{2} g \alpha \frac{\delta T}{\delta h} z^2 + g \alpha T_0 z. \tag{1.5.14}$$

An advantage of this formulation is that (1.5.11), (1.5.12), and (1.5.13) may be considered independently from (1.5.14) so that periodic boundary conditions may be applied to all the variables in all directions, including the vertical direction.

A quantity of central interest in thermal convection problems is the rate of heat transport by the fluid. In the Boussinesq equations the heat energy content of the fluid is proportional to the local temperature (the constant of proportionality being the specific heat) so the heat current is directly proportional to the "temperature current." This temperature current is defined by the temperature evolution equation (1.5.3) which, using the incompressibility condition, may be written

$$\frac{\partial T}{\partial t} + \nabla \cdot \mathbf{H} = 0, \tag{1.5.15}$$

where the current density $\mathbf{H}$ is

$$\mathbf{H} = \mathbf{u}T - \kappa \nabla T. \tag{1.5.16}$$

Given a unit vector $\mathbf{n}$ at position $\mathbf{x}$ at time t, $\mathbf{n} \cdot \mathbf{H}$ is (proportional to) the heat per unit time per unit area crossing an infinitesimal area element oriented normal to $\mathbf{n}$. For convection across a horizontal layer of thickness h with the top of the layer held at temperature T, the bottom held at $T + \delta T$, and either periodic or rigid insulating boundary conditions in the horizontal direction, the vertically averaged total heat flow is

$$h^{-1} \int dx \int dy \int dz \left[u_3 T - \kappa \frac{\partial T}{\partial z} \right] = h^{-1} \int dx \int dy \int u_3 T dz + \kappa h^{-1} \delta T A, \tag{1.5.17}$$

where $A = \int dx \int dy$ is the area of the horizontal cross section of the layer. The total heat flow is composed of two terms: the *conductive* heat flow ($\kappa \delta T A/h$) depends only on the boundary conditions, whereas the *convective* heat flow (the integral of $u_3 T$) is the heat transported by the flow field. A central problem of theoretical convection studies is to analyze and predict the convective heat transport for various geometries and imposed temperature differences.

1.6 References and further reading

The structure of the equations of motion of fluid mechanics are discussed in many texts, for example Tritton [1]. The incompressible Navier-Stokes and Boussinesq equations are determined from the Boltzmann equation of nonequilibrium statistical mechanics under the assumption that the flow field varies only on length and time scales much greater than the microscopic scales associated with the mean free path and mean free time of particles, and that velocities are much less than the speed of sound. This derivation provides explicit expressions for the macroscopic material parameters (viscosity, thermal conductivity, etc.) in terms of the microscopic interactions of the particles making up the fluid. See the recent work of Bardos, Golse, and Levermore [2].

Exercises

1 Derive the energy evolution equation (1.3.9).

2 Derive the vorticity evolution equation (1.3.15) from the incompressible Navier-Stokes equations. The following identities may be useful.

$$\nabla(\mathbf{A}\cdot\mathbf{B}) = \mathbf{A}\cdot\nabla\mathbf{B} + \mathbf{B}\cdot\nabla\mathbf{A} + \mathbf{A}\times\operatorname{curl}\mathbf{B} + \mathbf{B}\times\operatorname{curl}\mathbf{A}, \qquad \text{(E1.1)}$$

$$\operatorname{curl}(\mathbf{A}\times\mathbf{B}) = \mathbf{A}\operatorname{div}\mathbf{B} - \mathbf{B}\operatorname{div}\mathbf{A} + \mathbf{B}\cdot\nabla\mathbf{A} - \mathbf{A}\cdot\nabla\mathbf{B}. \qquad \text{(E1.2)}$$

3 In two dimensions the divergence-free condition on $\mathbf{u}$ may be imposed by writing

$$u_1 = -\frac{\partial\psi}{\partial y} \qquad u_2 = \frac{\partial\psi}{\partial x}, \qquad \text{(E1.3)}$$

introducing the stream function $\psi(\mathbf{x},t)$. Show that level sets of the stream function have tangent vectors parallel to $\mathbf{u}$, and find the relation between ψ and the scalar vorticity ω.

4 Show (1.4.11) follows from (1.4.10).

5 Derive the energy evolution equation (1.4.18).

6 Show that $\mathbf{P}\left\{\frac{d\mathbf{u}}{dt}\right\} = 0$ are the incompressible Euler equations (see footnote 1, page 6).

2

Dimensionless parameters and stability

2.1 Dimensionless parameters

Consider the incompressible Navier-Stokes equations

$$\frac{\partial \mathbf{u}}{\partial t} + \mathbf{u} \cdot \nabla \mathbf{u} + \frac{1}{\rho} \nabla p = \nu \Delta \mathbf{u} + \frac{1}{\rho} \mathbf{f}, \tag{2.1.1}$$

$$\nabla \cdot \mathbf{u} = 0 \tag{2.1.2}$$

with

$$\mathbf{x} \in [0, L]^d \tag{2.1.3}$$

and periodic boundary conditions. For a given functional form of the body force, the four parameters of the problem are the density ρ with dimensions mass per unit d-volume), the kinematic viscosity ν (with dimensions length^2 $\times$ time^{-1}), the system size L, and some normalization of the body force $\mathbf{f}$, such as the root mean square (r.m.s.) amplitude $L^{-d/2}\|\mathbf{f}\|_2$ (with dimensions mass $\times$ length $\times$ time^{-2} per unit d-volume).

This parameter set may be reduced by changing variables. Define the dimensionless length, time, velocity, and pressure variables

$$\mathbf{x}' = \mathbf{x}/L, \qquad t' = \nu t/L^2, \qquad \mathbf{u}' = \mathbf{u}L/\nu, \qquad p' = pL^2/\rho\nu^2, \tag{2.1.4}$$

and the dimensionless "unit-strength" body force

$$\mathbf{f}' = \frac{L^{d/2}}{\|\mathbf{f}\|_2}\mathbf{f}. \tag{2.1.5}$$

Then the Navier-Stokes equations become

$$\frac{\partial \mathbf{u}'}{\partial t'} + \mathbf{u}' \cdot \nabla' \mathbf{u}' + \nabla' p' = \Delta' \mathbf{u}' + \mathscr{G} \mathbf{f}' \tag{2.1.6}$$

$$\nabla' \cdot \mathbf{u}' = 0, \tag{2.1.7}$$

where

$$\mathbf{x}' \in [0,1]^d. \tag{2.1.8}$$

∇' and Δ' denote the gradient and Laplacian operators with respect to the primed variables. We have defined the Grashof number $\mathscr{G}$ by

$$\mathscr{G} = \frac{L^{3-d/2}}{\rho \nu^2} \|\mathbf{f}\|_2. \tag{2.1.9}$$

According to this simple rescaling then, the entire four parameter class of problems defined by (2.1.1)-(2.1.3) is equivalent to the one parameter family in (2.1.6)-(2.1.8). It is thus sufficient to consider the unit density, unit viscosity fluid on the unit d-torus, subsuming all the particular scales of the system into the amplitude of the applied body force. The single remaining parameter, $\mathscr{G}$, is the unique (up to powers) dimensionless measure of the ratio of the strength of the driving force to the damping coefficient. Problems with $\mathscr{G} \ll 1$ correspond to weakly forced or overdamped systems where we may anticipate relatively tame flows. On the other hand, if $\mathscr{G} \gg 1$, then the forcing is "strong" in an absolute sense, or the damping is truly weak, and more interesting dynamical behavior may be expected. This is an example of the power and convenience of the so-called scale invariance of the Navier-Stokes equations.

Another example is provided by the following discussion. Consider the Navier-Stokes equations

$$\frac{\partial \mathbf{u}}{\partial t} + \mathbf{u} \cdot \nabla \mathbf{u} + \frac{1}{\rho} \nabla p = \nu \Delta \mathbf{u} \tag{2.1.10}$$

$$\nabla \cdot \mathbf{u} = 0 \tag{2.1.11}$$

with

$$\mathbf{x} \in \Omega, \tag{2.1.12}$$

a d-dimensional set. Rather than driving the system with a body force we drive it by some boundary conditions,

$$\mathbf{u}|_{\partial\Omega} = \mathbf{U}(\mathbf{x}), \tag{2.1.13}$$

where $\mathbf{U}(\mathbf{x})$ is specified for $\mathbf{x} \in \partial\Omega$. Then there are again four parameters in the problem: ρ, ν, a length scale set by the system size, and a velocity scale set by the boundary conditions. (There is also a velocity scale set by the initial conditions which may dominate the flow at early times, an observation which applies to the body-forced example as well. But if the long-term properties of the flow are of interest and the effect of the initial data is transient, then presumably this scale will not be relevant.)

There are a number of possibilities for the system size length scale. A natural choice may be a length scale L based on the system volume, say,

$$L \sim \left(\int_\Omega d^d x \right)^{1/d}. \tag{2.1.14}$$

This scale does not take into account any of the geometrical features of Ω, however, and its utility is restricted to finite volume sets. Another length scale – the one we will use – is that defined by the spectrum of the Laplacian on Ω with Dirichlet boundary conditions. That is, suppose that Ω and $\partial\Omega$ are bounded enough and regular enough that the spectrum of the operator $-\Delta$, with vanishing boundary conditions on $\partial\Omega$, is discrete. Denote the eigenvalues by

$$0 < \lambda_1 \leq \lambda_2 \leq \lambda_3 \ldots. \tag{2.1.15}$$

These are real and positive because the negative Laplacian $-\Delta$ is a positive self-adjoint operator on these boundary conditions. They also have a dimension of $(\text{length})^{-2}$, so the largest length scale based on the spectrum is

$$L \sim \frac{1}{\sqrt{\lambda_1}}. \tag{2.1.16}$$

The proportionality factor, an $O(1)$ pure number, is chosen for convenience. For example, a factor $\pi\sqrt{d}$ might be used so that this length scale for the hypercube $[0, L]^d$ is precisely L.

Likewise, there is some freedom in our choice of velocity scales based on the boundary conditions. We'll use the r.m.s. velocity on the boundary to define the velocity scale U:

$$U^2 = \frac{\int_{\partial\Omega} |\mathbf{U}|^2 \, d^{d-1}x}{\int_{\partial\Omega} d^{d-1}x}. \tag{2.1.17}$$

The set of parameters in this problem may now be reduced, again by changing variables. Define the dimensionless length, time, velocity, and pressure variables

$$\mathbf{x}' = \mathbf{x}/L, \qquad t' = Ut/L, \qquad \mathbf{u}' = \mathbf{u}/U, \qquad p' = p/\rho U^2. \tag{2.1.18}$$

In these variables, the Navier-Stokes equations (2.1.10) to (2.1.12) become

$$\frac{\partial \mathbf{u}'}{\partial t'} + \mathbf{u}' \cdot \nabla' \mathbf{u}' + \nabla' p' = \frac{1}{R} \Delta' \mathbf{u}', \tag{2.1.19}$$

$$\nabla' \cdot \mathbf{u}' = 0 \tag{2.1.20}$$

with

$$\mathbf{x}' \in \Omega', \tag{2.1.21}$$

a subset of $\mathbf{R}^d$ for which the smallest eigenvalue of the negative Laplacian with vanishing boundary conditions $\partial\Omega'$ is O(1). The system parameters are all subsumed in the single dimensionless parameter called the *Reynolds number*

$$R = \frac{UL}{\nu}. \tag{2.1.22}$$

The boundary conditions for $\mathbf{u}'$ are

$$\mathbf{u}'|_{\partial\Omega} = U^{-1}\mathbf{U}, \tag{2.1.23}$$

so that the boundary is driven at unit (r.m.s.) speed. The Reynolds number is a measure of the ratio of the imposed velocity scale U to the "viscous" velocity scale determined by the system size L and the momentum diffusion time L^2/ν. Alternatively it may be thought of as a ratio of the driving from the boundary to the damping from the coefficient of viscosity. Low Reynolds numbers mean strong momentum diffusion – alternatively, weak driving or strong damping – suggesting dynamically constrained flows. High Reynolds numbers correspond to relatively strongly driven, underdamped systems. Indeed, the singular limit $R \to \infty$ transforms the Navier-Stokes equations into the conservative Euler's equations. Solutions of the Navier-Stokes equations at high Reynolds numbers may appear, locally, similar to inviscid flows solving Euler's equations. The boundary conditions for the Navier-Stokes equations are fundamentally different from those for Euler's equations, however, and viscous boundary layers are found near rigid walls.[1] This effect of viscosity is a fundamental source of the difference between solutions of the high Reynolds number Navier-Stokes equations and Euler's equations.

The Reynolds number based on the initial data is the relevant dimensionless parameter for free decay problems. Consider, for example, the force-free Navier-Stokes equations (2.1.10) to (2.1.12) with either periodic or stationary rigid boundaries. Then the only velocity scale is that provided by the initial condition $\mathbf{u}_0(\mathbf{x}) = \mathbf{u}(\mathbf{x}, 0)$. Defining the velocity scale

[1] This topic will arise again in Chapter 10.

by the r.m.s. value of the initial disturbance,

$$U^2 = \frac{\int_\Omega |\mathbf{u}_0|^2 \, d^d x}{\int_\Omega d^d x}, \tag{2.1.24}$$

and the Reynolds number, as before, by $R = UL/\nu$, the change of variables in (2.1.18) leads to the same dimensionless form of the Navier-Stokes equations as in (2.1.19) to (2.1.21). This class of problem is then equivalent to the unit-density, unit-viscosity fluid in a "unit-eigenvalue" volume with unit-norm initial data. The energy of the initial flow configuration, the system size, and the damping coefficient are all combined into the single Reynolds number. For such a problem with small Reynolds number, overdamped dynamics, characterized by a monotonic relaxation to equilibrium, is expected. In high Reynolds number cases, however, the system is highly underdamped and some kind of overshoot would not be surprising. As we will see in later chapters, the rigorous estimates available for solutions of the Navier-Stokes equations are consistent with this picture. The conclusion is that the scale invariance of the Navier-Stokes equations tells us that an increase in the energy density of the initial condition (or the velocity on the boundary) is equivalent to a decrease in the fluid viscosity, either of which is equivalent to increasing the system size.

In situations involving the Reynolds number there is an alternative choice of rescaling which illustrates another aspect of the difference between $R \ll 1$ and $R \gg 1$. Given the system size scale L and velocity scale U, define the dimensionless variables

$$\mathbf{x}'' = \mathbf{x}/L, \qquad t'' = \nu t/L^2, \qquad \mathbf{u}'' = \mathbf{u}/U, \qquad p'' = pL/\rho\nu U. \tag{2.1.25}$$

In these variables, the Navier-Stokes equations (2.1.10) and (2.1.11) become

$$\frac{\partial \mathbf{u}''}{\partial t''} + R\,\mathbf{u}'' \cdot \nabla''\mathbf{u}'' + \nabla'' p'' = \Delta''\mathbf{u}'' \qquad \nabla'' \cdot \mathbf{u}'' = 0, \tag{2.1.26}$$

and R is now the coefficient of the nonlinear term in the momentum equation. In the $R \to 0$ limit, (2.1.26) linearizes to a diffusion equation for the divergence-free velocity vector field. The associated dynamics is overdamped in the sense that inertia plays no role in the fluid motion. In the other extreme the nonlinear "inertial" term plays a dominant role in the dynamics. The Reynolds number may thus be interpreted as a ratio of the strength of inertial forces to viscous forces.

There are dimensionless formulations of the thermal convection problems described by the Boussinesq equations in section 1.5,

$$\frac{\partial \mathbf{u}}{\partial t} + \mathbf{u} \cdot \nabla \mathbf{u} + \frac{1}{\rho} \nabla p = \nu \Delta \mathbf{u} + \frac{1}{\rho} \hat{\mathbf{k}} g \alpha T, \tag{2.1.27}$$

$$\nabla \cdot \mathbf{u} = 0, \tag{2.1.28}$$

$$\frac{\partial T}{\partial t} + \mathbf{u} \cdot \nabla T = \kappa \Delta T. \tag{2.1.29}$$

For any of the formulations described in section 1.5, we may choose the length scale determined by the vertical gap height h, the associated time scale defined by the thermal diffusion coefficient κ, and the temperature scale imposed by the difference δT. That is, in the dimensionless variables

$$\begin{aligned} \mathbf{x}' &= \mathbf{x}/h, \qquad t' = \kappa t/h^2, \qquad \mathbf{u}' = \mathbf{u}h/\kappa, \\ p' &= ph^2/\rho\kappa^2 \qquad \text{and} \qquad T' = T/\delta T, \end{aligned} \tag{2.1.30}$$

the dynamical equations become

$$\frac{\partial \mathbf{u}'}{\partial t'} + \mathbf{u}' \cdot \nabla' \mathbf{u}' + \nabla' p' = \sigma \Delta' \mathbf{u}' + \hat{\mathbf{k}} \sigma Ra\, T', \tag{2.1.31}$$

$$\nabla' \cdot \mathbf{u}' = 0 \tag{2.1.32}$$

$$\frac{\partial T'}{\partial t'} + \mathbf{u}' \cdot \nabla' T' = \Delta' T', \tag{2.1.33}$$

for a unit temperature difference across a unit height gap, where we have defined the dimensionless *Prandtl number*

$$\sigma = \frac{\nu}{\kappa}, \tag{2.1.34}$$

and the *Rayleigh number*

$$Ra = \frac{\alpha g \delta T h^3}{\rho \nu \kappa}. \tag{2.1.35}$$

The Rayleigh number is a measure of the ratio of the driving (the imposed average temperature gradient) to the damping (the viscosity and the thermal diffusion) in the system. A small Rayleigh number indicates a relatively weakly driven, highly constrained system while a large Rayleigh number, with its increased coupling between the temperature and flow fields, allows for more interesting dynamical behavior. The Prandtl number σ, the ratio of the viscosity to the thermal diffusion coefficient, is a ratio of material parameters and as such is a characteristic of the fluid itself rather than of the constraints imposed on the flow field. It does not involve extensive variables comprising the driving or damping, and so plays a less transparent role in the dynamics. The formal limits of infinite or vanishing Prandtl number, however, reduce the equations to models which are of interest in their own right.

A quantity of central interest in thermal convection studies is the heat flow. The natural dimensionless measure of the heat flow is the ratio of the total time averaged rate of heat transport to the conductive rate of heat transport (recall (1.5.17)), known as the *Nusselt number*

$$Nu = 1 + \frac{h}{\kappa \delta T} \langle \overline{u_3 T} \rangle = 1 + \langle \overline{u_3' T'} \rangle, \tag{2.1.36}$$

where the overbar means spatial average and $\langle \cdot \rangle$ means time average. In the purely conductive state where the flow field vanishes, the Nusselt number is 1. A thermally stimulated flow typically enhances the heat flow, increasing Nu. The relationship between the imposed temperature difference and the heat flow, for a given geometry, is contained in the dimensionless Nusselt number as a function of the Rayleigh and Prandtl numbers:

$$Nu = Nu(Ra, \sigma). \tag{2.1.37}$$

A central problem of theoretical convection studies is to deduce this relationship from the Boussinesq equations of motion.

2.2 Linear and nonlinear stability, differential inequalities

Suppose $\mathbf{U}(\mathbf{x}), P(\mathbf{x})$ solve the stationary Navier-Stokes equations in a region Ω with some specified stationary boundary conditions:

$$\mathbf{U} \cdot \nabla \mathbf{U} + \frac{1}{\rho} \nabla P = \nu \Delta \mathbf{U}, \tag{2.2.1}$$

$$\nabla \cdot \mathbf{U} = 0. \tag{2.2.2}$$

We will refer to such time independent solutions as stationary solutions or "fixed points," because they are fixed points of the system dynamics in velocity-pressure configuration space. The relevance of a stationary solution to "real" problems depends in large part on its stability properties. If infinitesimal perturbations of a stationary solution are amplified with time – that is, if the fixed point is an unstable fixed point – then the evolution naturally drives the system away from that stationary state. A real system is always subject to some small fluctuations which may provide such perturbations. If, however, perturbations from the stationary solution decay with time – that is, if the fixed point is a stable fixed point – then the solution is a candidate time-asymptotic solution under realistic constraints.

In order to test for stability or instability, we may consider the evolution of deviations from the stationary state. Let an arbitrary solution $\mathbf{u}(\mathbf{x}, t)$, $p(\mathbf{x}, t)$ be decomposed according to

$$\begin{aligned} \mathbf{u}(\mathbf{x}, t) &= \mathbf{U}(\mathbf{x}) + \delta\mathbf{u}(\mathbf{x}, t) \\ p(\mathbf{x}, t) &= P(\mathbf{x}) + \delta p(\mathbf{x}, t). \end{aligned} \tag{2.2.3}$$

Then the perturbations $\delta\mathbf{u}$ and δp satisfy

$$\frac{\partial \delta\mathbf{u}}{\partial t} + \delta\mathbf{u} \cdot \nabla\delta\mathbf{u} + \delta\mathbf{u} \cdot \nabla\mathbf{U} + \mathbf{U} \cdot \nabla\delta\mathbf{u} + \frac{1}{\rho}\nabla\delta p = \nu\Delta\delta\mathbf{u}, \tag{2.2.4}$$

$$\nabla \cdot \delta\mathbf{u} = 0 \tag{2.2.5}$$

along with the appropriate boundary conditions. (For example, if the boundary conditions for $\mathbf{U}$ and P are periodic, then they are also periodic for the deviations; if $\mathbf{U}$ is specified on a boundary, then $\delta\mathbf{u}$ vanishes there, etc.) The aim of stability theory is to extract information about the evolution of the deviations without solving the full nonlinear initial-boundary value problem in (2.2.4) and (2.2.5).

The "linear stability" of a stationary solution is, by definition, determined by the evolution of infinitesimal perturbations. For infinitesimal perturbations, the quadratic $\delta\mathbf{u} \cdot \nabla\delta\mathbf{u}$ term is negligible (of higher order) and so may be dropped. Then the system's evolution may be linearized to

$$\frac{\partial \delta\mathbf{u}}{\partial t} + \delta\mathbf{u} \cdot \nabla\mathbf{U} + \mathbf{U} \cdot \nabla\delta\mathbf{u} + \frac{1}{\rho}\nabla\delta p = \nu\Delta\delta\mathbf{u}, \tag{2.2.6}$$

$$\nabla \cdot \delta\mathbf{u} = 0. \tag{2.2.7}$$

The time evolution of this linear (albeit generically nonconstant coefficient) system of equations may be reduced to an eigenvalue problem by imposing an $e^{-\lambda t}$ time dependence for solutions, and superposing to obtain the general solution. The eigenvalue problem for a solution of the form $\exp(-\lambda t)\delta\mathbf{u}(\mathbf{x})$ is

$$\lambda\delta\mathbf{u} = -\nu\Delta\delta\mathbf{u} + \delta\mathbf{u} \cdot \nabla\mathbf{U} + \mathbf{U} \cdot \nabla\delta\mathbf{u} + \frac{1}{\rho}\nabla\delta p \tag{2.2.8}$$

$$\nabla \cdot \delta\mathbf{u} = 0 \tag{2.2.9}$$

with the associated boundary conditions: $\delta\mathbf{u}$ vanishing at rigid boundaries, or periodic as appropriate. The operator whose spectrum is of interest in (2.2.8) and (2.2.9) is not generally self-adjoint, so the eigenvalues typically have both real and imaginary parts. If the real parts are all positive, then the amplitude of every infinitesimal perturbation decays exponentially with time. The underlying stationary solution is then said to be "linearly stable." If the largest real part of any eigenvalue is exactly zero, then the stationary solution is said to be "marginal." Note that linear stability does not guarantee that *any* finite perturbation decays. Linear stability is a relatively weak notion, and quite distinct from some more robust stability conditions relevant to finite amplitude perturbations. If at least one eigenvalue has a positive real part, however, then some infinitesimal perturbation grows exponentially. The associated fixed point is then "linearly unstable." This is a strong notion of instability: It guarantees not only that there is some deviation which will grow, but also that it will be amplified exponentially in time. (A small perturbation growing exponentially will eventually grow large enough for the neglected nonlinear terms to come into play, and then the linearized evolution equation no longer applies.) Linear instability is a sufficient condition for instability, but linear or marginal stability is only a necessary condition for stability.

To establish stability against any finite amplitude perturbations, no matter how small, the full nonlinear equations must be used. We may proceed, however, by noting that a sufficient condition for stability is that the kinetic energy in the deviations should eventually vanish. Dotting $\delta\mathbf{u}$ into (2.2.4), integrating, and integrating by parts using both the divergence free and boundary conditions, we see that the L^2 norm of a perturbation evolves according to

$$\frac{d}{dt}\left(\frac{1}{2}\|\delta\mathbf{u}\|_2^2\right) = -\nu\|\nabla\delta\mathbf{u}\|_2^2 - \int_\Omega \delta\mathbf{u}\cdot(\nabla\mathbf{U})\cdot\delta\mathbf{u}\,d^dx. \tag{2.2.10}$$

Clearly the viscosity works to inhibit a perturbation, but the influence of the base solution is more subtle. In order to establish the "nonlinear stability" of $\mathbf{U}$, it is sufficient to show that the L^2 norm of any relevant perturbation vanishes as $t \to \infty$. To show this, it is sufficient to show that the L^2 norm of any relevant perturbation vanishes exponentially. In particular, this is the case if the right hand side of (2.2.10) above is negative and less than or equal to $-c\|\delta\mathbf{u}\|_2^2$ for some positive constant c uniformly for *all* relevant flow fields $\delta\mathbf{u}$.

This last assertion follows from Gronwall's inequality, a very useful example of the power and utility of differential inequalities. We state and prove it as

Lemma 2.1 *(Gronwall's lemma): Suppose $f(t)$ is a (real) function whose derivative is bounded according to*

$$\frac{df}{dt} \leq g(t)f + h(t) \tag{2.2.11}$$

for some (real) functions $g(t)$ and $h(t)$. Then $f(t)$ is bounded pointwise in time according to

$$f(t) \leq f(0)\exp[G(t)] + \int_0^t \exp[G(t-s)]\, h(s)\, ds, \tag{2.2.12}$$

where $f(0)$ is the value of f at $t = 0$ and

$$G(t) = \int_0^t g(r)\, dr. \tag{2.2.13}$$

Proof: Note first that

$$\frac{df}{dt} - g(t)f = \exp[G]\, \frac{d}{dt}\left(\exp[-G]\, f\right). \tag{2.2.14}$$

Thus, rearranging (2.2.11), using (2.2.14) and multiplying through by $\exp[-G]$ (which is manifestly positive and so doesn't affect the direction of the inequality),

$$\frac{d}{dt}\left(\exp[-G]f\right) \leq \exp[-G]\, h. \tag{2.2.15}$$

Now integrate over the time variable from 0 to t without violating the inequality. The left-hand side is a perfect derivative, so

$$\exp[-G(t)]f(t) - \exp[-G(0)]f(0) \leq \int_0^t \exp[-G(s)]\, h(s)\, ds. \tag{2.2.16}$$

Then the advertised result in (2.2.12) follows from this by rearranging, multiplying through by $\exp[G]$, and noting that $G(0) = 0$ and $G(t) - G(s) = G(t-s)$. □

Returning to the nonlinear stability problem, suppose that the stationary solution $\mathbf{U}(\mathbf{x})$ and the boundary conditions are such that there exists a constant $c > 0$ so that

$$c\, \|\mathbf{v}\|_2^2 \leq \nu \|\nabla\mathbf{v}\|_2^2 + \int_\Omega \mathbf{v} \cdot (\nabla\mathbf{U}) \cdot \mathbf{v}\, d^d x \tag{2.2.17}$$

for *every* divergence-free velocity vector field satisfying the perturbation's boundary conditions. Then the L^2 norm of any perturbation would satisfy the differential inequality

$$\frac{d}{dt}\left(\frac{1}{2}\|\delta\mathbf{u}\|_2^2\right) \leq -c\,\|\delta\mathbf{u}\|_2^2, \tag{2.2.18}$$

and by Gronwall's lemma (with $g(t) = -c$ and $h(t) = 0$) we find

$$\|\delta\mathbf{u}(\cdot,t)\|_2^2 \leq \|\delta\mathbf{u}(\cdot,0)\|_2^2 \exp(-ct). \tag{2.2.19}$$

Hence any finite amplitude, finite energy perturbation would decay and the underlying base solution would be nonlinearly stable. The condition (2.2.17) is sufficient for nonlinear stability, but as already discussed, direct confirmation of this is not very practical. This nonlinear stability problem can be reformulated, however, as a linear spectral problem along the lines of the linear stability analysis. This fact, which is peculiar to these kinds of hydrodynamic stability problems, is possible due to the fact that the right-hand side of the "energy" evolution equation (2.2.10) is quadratic in $\delta\mathbf{u}$.

In order to check for the inequality in (2.2.17), one needs to evaluate or estimate the quantity

$$\inf\left(\frac{\nu\|\nabla\mathbf{v}\|_2^2 + \int_\Omega \mathbf{v}\cdot(\nabla\mathbf{U})\cdot\mathbf{v}\,d^dx}{\|\mathbf{v}\|_2^2}\right), \tag{2.2.20}$$

where the infimum is taken over all divergence-free vector fields satisfying the perturbation's boundary conditions. If the smallest value of the ratio is positive then it plays the role of c, the slowest possible decay rate of a perturbation. Because of the homogeneity of the ratio (it is unaffected by the rescaling $\mathbf{v} \to \alpha\mathbf{v}$), the value of the infimum is the same as the infimum of the numerator alone over all divergence-free, unit L^2 norm vector fields satisfying the boundary conditions.

Minimizing the numerator in (2.2.20), subject to the constraints $\nabla\cdot\mathbf{v} = 0$ and $\|\mathbf{v}\|_2 = 1$, may be accomplished by the calculus of variations. The minimum is realized by a field satisfying the associated Euler-Lagrange equations. In component notation, the functional to be minimized is

$$F\{\mathbf{v}\} = \int\left(\nu\frac{\partial v_i}{\partial x_j}\frac{\partial v_i}{\partial x_j} + \frac{1}{2}v_i\left[\frac{\partial U_i}{\partial x_j} + \frac{\partial U_j}{\partial x_i}\right]v_j - 2p\frac{\partial v_i}{\partial x_i} - \lambda v_i v_i\right)d^dx, \tag{2.2.21}$$

where $p = p(\mathbf{x})$ is the Lagrange multiplier for the (local) divergence-free constraint and λ is the Lagrange multiplier for the (global) normalization constraint. The minimizing field satisfies the Euler-Lagrange equations

$$0 = \frac{\delta F}{\delta v_i} = 2\left(-\nu\Delta v_i + D_{ij}v_j + \frac{\partial p}{\partial x_i} - \lambda v_i\right), \tag{2.2.22}$$

with the symmetric matrix multiplication operator

$$D_{ij}(\mathbf{x}) = \frac{1}{2}\left(\frac{\partial U_i}{\partial x_j} + \frac{\partial U_j}{\partial x_i}\right). \tag{2.2.23}$$

That is, the minimizing field is a solution of the eigenvalue problem

$$\lambda v_i = -\nu\Delta v_i + D_{ij}v_j + \frac{1}{\rho}\frac{\partial p}{\partial x_i}, \tag{2.2.24}$$

$$\frac{\partial v_i}{\partial x_i} = 0. \tag{2.2.25}$$

(The Lagrange multiplier enforcing the vanishing divergence condition plays the role of a "pressure.") An acceptable solution to this eigenvalue problem is a unit-normalizable (in L^2) vector field satisfying the boundary conditions, along with its associated eigenvalue. In contrast to the eigenvalue problem for the linearized stability problem, the nonlinear stability question generally leads to a self-adjoint operator, and thus a real spectrum. The stability issue boils down to the question of the sign of the smallest eigenvalue for this self-adjoint operator: If the bottom of the spectrum is positive, then all deviations from the stationary solution decay exponentially in time. Transients decay at a minimum rate given by the lowest eigenvalue.

The conclusion of these considerations is that sufficient conditions for stability or instability may be obtained from the spectral analysis of some linear operators. Linear stability analysis, usually associated with the spectrum of a nonself-adjoint operator, can provide sufficient conditions for instability. Nonlinear stability analysis – also known as the "energy method" – can provide sufficient conditions for stability through the analysis of an eigenvalue problem for a self-adjoint operator.

Much of the work of hydrodynamic stability theory, both linear and nonlinear, is concerned with determining the precise conditions under which a few-parameter family of stationary solutions of the Navier-Stokes equations is stable. Stability or instability generally depends on the values of these parameters. The Navier-Stokes equations' scale invariance, however, limits the number of combinations of variables which are relevant to stability questions. For example, in the general setting under consideration in this section we can see how an appropriately defined

Reynolds number can play the key role in stability considerations. We restate the problem in dimensionless variables defined by the length scale based on the spectrum of the negative Laplacian on divergence-free vector fields in the spatial region of interest with the perturbation's boundary conditions (which we assume is bounded away from zero),

$$L = \frac{1}{\sqrt{\lambda_1}}, \tag{2.2.26}$$

the time scale L^2/ν, and the velocity scale

$$U = \sup_{\mathbf{x}} |\mathbf{U}(\mathbf{x})|. \tag{2.2.27}$$

In terms of the dimensionless variables

$$\mathbf{x}' = \mathbf{x}/L, \qquad t' = \nu t/L^2, \qquad \mathbf{u}' = \mathbf{u}/U, \qquad p' = pL/\rho\nu U, \tag{2.2.28}$$

the deviations' evolution equations (2.2.4) and (2.2.5) become

$$\frac{\partial \delta\mathbf{u}'}{\partial t'} + R\left(\delta\mathbf{u}' \cdot \nabla'\delta\mathbf{u}' + \delta\mathbf{u}' \cdot \nabla'\mathbf{U}' + \mathbf{U}' \cdot \nabla'\delta\mathbf{u}'\right) + \frac{1}{\rho}\nabla'\delta p' = \Delta'\delta\mathbf{u}', \tag{2.2.29}$$

$$\nabla' \cdot \delta\mathbf{u}' = 0, \tag{2.2.30}$$

where the Reynolds number is defined as

$$R = \frac{UL}{\nu}. \tag{2.2.31}$$

The Reynolds number is the only remaining parameter in the problem. Then the nonlinear stability problem is determined by the sign of

$$\inf\left(\frac{\|\nabla\mathbf{v}\|_2^2 + R\int_{\Omega'} \mathbf{v} \cdot (\nabla'\mathbf{U}') \cdot \mathbf{v}\, d^dx'}{\|\mathbf{v}\|_2^2}\right), \tag{2.2.32}$$

the infimum being taken, as before, over divergenceless vector fields $\mathbf{v}$ satisfying the boundary conditions. Integrating by parts and using the vanishing divergence of $\mathbf{v}$, the second term in the numerator above is bounded from below according to

$$\begin{aligned} R\int_{\Omega'} v_i v_j \frac{\partial U_i'}{\partial x_j'}\, d^dx' &= -R\int_{\Omega'} U_i' v_j \frac{\partial v_i}{\partial x_j'}\, d^dx' \\ &\geq -R \sup_{\mathbf{x}} |\mathbf{U}'(\mathbf{x})| \int_{\Omega'} |\mathbf{v}||\nabla'\mathbf{v}|\, d^dx' \\ &\geq -R\,\|\mathbf{v}\|_2\,\|\nabla'\mathbf{v}\|_2 \\ &\geq -\frac{1}{2}R\left(R\|\mathbf{v}\|_2^2 + \frac{1}{R}\|\nabla'\mathbf{v}\|_2^2\right). \end{aligned} \tag{2.2.33}$$

In the above we have used the following:

- $|\mathbf{U}'|$ has a supremum of 1 in these velocity units,
- Cauchy's inequality $|\int fg|^2 \le \int |f|^2 \int |g|^2$
- any real numbers a, b and a positive number c satisfy $2ab \le (ca^2 + b^2/c)$. (We have chosen $c = R$.)

Inserting (2.2.33) into (2.2.32), we find

$$\inf\left(\frac{\|\nabla \mathbf{v}\|_2^2 + R\int_{\Omega'} \mathbf{v}\cdot(\nabla'\mathbf{U}')\cdot \mathbf{v}\, d^dx'}{\|\mathbf{v}\|_2^2}\right) \ge \frac{1}{2}\inf\left(\frac{\|\nabla \mathbf{v}\|_2^2}{\|\mathbf{v}\|_2^2} - R^2\right). \tag{2.2.34}$$

The sufficient condition for stability holds if the Reynolds number is small enough, specifically if

$$R^2 \le \inf \frac{\|\nabla \mathbf{v}\|_2^2}{\|\mathbf{v}\|_2^2}. \tag{2.2.35}$$

The L^2 norm of the gradient of a function and the L^2 norm of the function itself are related by Poincaré's inequality which we state and prove as

Theorem 2.1 *(Poincaré's inequality): Suppose Ω is a set for which the negative Laplacian, $-\Delta$, along with boundary conditions, is a strictly positive self-adjoint linear operator with a discrete spectrum and smallest eigenvalue $\lambda_1 > 0$. Suppose further that $f(\mathbf{x})$ and its gradient $\nabla f(\mathbf{x})$ are square integrable on a set Ω, that $f(\mathbf{x})$ satisfies the boundary conditions, and that the boundary conditions allow for the integration by parts*

$$\int_\Omega f(\mathbf{x})[-\Delta f(\mathbf{x})]\, d^dx = \int_\Omega |\nabla f(\mathbf{x})|^2\, d^dx. \tag{2.2.36}$$

Then

$$\|f\|_2^2 \le \frac{1}{\lambda_1}\|\nabla f\|_2^2. \tag{2.2.37}$$

Proof: Let $\phi_1(\mathbf{x}), \phi_2(\mathbf{x}), \ldots$ be the complete orthonormal basis of eigenfunctions of $-\Delta$ with the boundary conditions on Ω, corresponding respectively to the eigenvalues $0 < \lambda_1 \le \lambda_2 \le \ldots$. Then

$$f(\mathbf{x}) = \sum_{n=1}^{\infty} f_n \phi_n(\mathbf{x}), \tag{2.2.38}$$

where

$$f_n = \int_\Omega \phi_n^*(\mathbf{x}) f(\mathbf{x})\, d^d x. \tag{2.2.39}$$

According to Parseval's theorem,

$$\|f\|_2^2 = \sum_{n=1}^{\infty} |f_n|^2 \tag{2.2.40}$$

and

$$\|\nabla f\|_2^2 = \sum_{n=1}^{\infty} \lambda_n |f_n|^2. \tag{2.2.41}$$

Thus, because $\lambda_n/\lambda_1 \geq 1$ for all n,

$$\|f\|_2^2 = \sum_{n=1}^{\infty} |f_n|^2 \leq \sum_{n=1}^{\infty} \frac{\lambda_n}{\lambda_1} |f_n|^2 = \frac{1}{\lambda_1} \|\nabla f\|_2^2. \tag{2.2.42}$$

We note that although this proof has been phrased in terms of a scalar function $f(\mathbf{x})$, an identical argument holds for the Laplacian restricted to divergence-free vector fields $\mathbf{v}(\mathbf{x})$. □

For the problem at hand, then, the ratio of the L^2 norm of the gradient of $\mathbf{v}$ and the L^2 norm of $\mathbf{v}$ is bounded from below by the lowest eigenvalue of $-\Delta$. In the dimensionless length units defined by (2.2.28), though, this smallest eigenvalue is precisely 1. Thus, using some assumptions on the spectrum of the Laplacian in the region and boundary conditions of interest, we have very generally established the nonlinear stability (and, essentially, the uniqueness) of steady solutions of the Navier-Stokes equations under the restriction

$$R < 1. \tag{2.2.43}$$

This result illustrates the general characteristic of stability problems in fluid dynamics: steady "laminar" solutions are often realized for small Reynolds numbers (or Rayleigh or Grashof numbers). These relatively simple flow fields may lose their stability – sufficient conditions for which may be established by linear stability analysis – at some higher "critical" value R_c^{lin} of the Reynolds number. For small Reynolds numbers the spectrum of the linearized evolution operator will be dominated by the Laplacian, with its nonpositive real part, and the critical Reynolds number R_c^{lin} is defined as the smallest value of R where the real part of an eigenvalue crosses zero. Then a bifurcation from the steady state occurs and other methods must be used to analyze, characterize, and classify typical solutions. Linear stability theory may, however, fail to predict a loss of stability; it is possible that the real parts of the eigenvalues of the linearized operator remain negative at all Reynolds numbers (i.e., $R_c^{lin} = \infty$). This does not mean that the underlying solution remains stable to finite amplitude perturbations, though. In this case the only information to be gleaned from these kinds of analyses is the possibility that finite amplitude perturbations may not decay if the nonlinear stability criterion fails above another critical Reynolds number, R_c^{nonlin}. Because nonlinear stability implies linear stability, $R_c^{nonlin} \leq R_c^{lin}$. For finite R_c^{nonlin}, a strict inequality $R_c^{nonlin} < R_c^{lin}$ means that a subcritical bifurcation from the steady state is possible. Equality of the linear and nonlinear critical Reynolds numbers yields the strongest result: $R_c^{nonlin} = R_c^{lin} = R_c < \infty$ implies a supercritical bifurcation precisely at the critical Reynolds number R_c. (An example of this is developed in the exercises.) The nature and stability of the bifurcating solution require their own analysis, for example, via the asymptotic methods of amplitude equations.

2.3 References and further reading

Linear and nonlinear stability theory is discussed in detail in Drazin and Reid [3]. The energy method for nonlinear stability is developed for a variety of problems in Straughan [4]. Linear stability can, in some cases, be elevated to "local" nonlinear stability, and these methods have been used to identify stable solutions of the Navier-Stokes equations from solutions of approximate problems (see Titi [5]). Stability criteria can also be applied to time periodic solutions (see Titi [6]).

Exercises

1 Verify the rescalings that lead to the nondimensional formulations in (2.1.6), (2.1.19), (2.1.26), and (2.1.31).

2 Consider the problem of $2d$ convection, where z is the vertical direction and x is the horizontal direction, with stress-free boundaries on top and bottom and periodic boundary conditions on $[0, L]$ horizontally. Compute the critical Rayleigh number of linear stability analysis about the pure conduction state, and show that its minimum possible value is $27\pi^4/4$.

3 Consider the nonlinear stability of the pure conduction state of the 2*d* convection problem outlined in the previous exercise. Show that the energy method leads to the same spectral problem as for the linearized stability analysis. Discuss the nature of the bifurcation at the critical Rayleigh number.

4 Show that the velocity field, known as Couette flow,

$$\mathbf{U}(\mathbf{x}) = \hat{\mathbf{i}} Uz/h, \tag{E2.1}$$

is an exact stationary solution of the Navier-Stokes equations corresponding to a fluid sheared between a stationary plate at $z = 0$ and one moving with speed U in the x-direction at $z = h$. Imposing periodic conditions on $[0, L_1]$ and $[0, L_2]$ in the x- and y-directions respectively, reduce the linear stability analysis to ordinary differential equations for the eigenvalue problem. Can you solve them?

5 Set up the nonlinear stability analysis for Couette flow as described in the previous exercise. From the integral form in (2.2.17) use Poincaré's inequality to derive a sufficient condition for nonlinear stability in terms of the Reynolds number

$$R = \frac{Uh}{\nu}. \tag{E2.2}$$

6 Suppose $f(x)$ is a square integrable periodic function on $[0, L]$, with a square integrable derivative. Show that Poincaré's inequality,

$$\|f\|_2^2 \leq \left(\frac{L}{2\pi}\right)^2 \left\|\frac{df}{dx}\right\|_2^2, \tag{E2.3}$$

holds iff $f(x)$ has mean zero, i.e., $\int_0^L f(x)\,dx = 0$.

3
Turbulence

3.1 Introduction

Turbulent motion in fluids is a familiar phenomenon from our everyday experience, but it is nevertheless an extremely difficult thing to define quantitatively. For the most part, the best that can be done to define turbulence is to list some of its characteristics: It is unsteady chaotic flow, apparently random, with fluid motions distributed over a relatively wide range of length and time scales. The complicated spatio-temporal structure of turbulent velocity fields renders their analytical description impossible, and the large number of degrees of freedom and the wide range of scales in turbulent flows result in difficult problems for numerical analysis, taxing both the speed and memory capacities of present day computers.

3.2 Statistical turbulence theory and the closure problem

Because of the effectively random behavior of turbulent flows, it is natural to attempt a statistical formulation. This is the classical approach to turbulence theory. The idea is to decompose a turbulent velocity vector field into a mean and a fluctuating part in an attempt to extract the relevant mean physical quantities. The "mean" in this approach may be a time average – appropriate for a steady configuration which, although fluctuating at all times, has well behaved time averaged characteristics – or an ensemble average where the average is over initial conditions in some class. To illustrate this decomposition we will consider only a steady state turbulent flow, assuming that well defined time averages of all quantities exist.

Suppose that $\mathbf{u}(\mathbf{x},t)$ is a turbulent solution of the Navier-Stokes equations

$$\frac{\partial \mathbf{u}}{\partial t} + \mathbf{u} \cdot \nabla \mathbf{u} + \frac{1}{\rho} \nabla p = \nu \Delta \mathbf{u} + \frac{1}{\rho} \mathbf{f}, \tag{3.2.1}$$

$$\nabla \cdot \mathbf{u} = 0, \tag{3.2.2}$$

with a time-independent body force $\mathbf{f}(\mathbf{x})$ for $\mathbf{x} \in \Omega$ with some specified time independent boundary conditions. We decompose $\mathbf{u}(\mathbf{x},t)$ as

$$\mathbf{u}(\mathbf{x},t) = \mathbf{U}(\mathbf{x}) + \mathbf{v}(\mathbf{x},t), \tag{3.2.3}$$

where $\mathbf{U}$ is the time averaged velocity field,

$$\mathbf{U}(\mathbf{x}) = \langle \mathbf{u}(\mathbf{x},\cdot) \rangle = \lim_{\tau \to \infty} \frac{1}{\tau} \int_0^\tau \mathbf{u}(\mathbf{x},t)\,dt, \tag{3.2.4}$$

and $\mathbf{v}(\mathbf{x},t)$ is the time-dependent fluctuating component satisfying

$$\langle \mathbf{v}(\mathbf{x},t) \rangle = 0. \tag{3.2.5}$$

The mean $\mathbf{U}$ satisfies the time independent boundary conditions that $\mathbf{u}$ satisfies, while the fluctuation $\mathbf{v}$ satisfies the homogeneous version of the boundary conditions. That is, if a component of $\mathbf{u}$ is specified on the boundary, then the corresponding component of $\mathbf{v}$ vanishes there. If the normal derivative of a component of $\mathbf{u}$ is given, then the normal derivative of that component of $\mathbf{v}$ vanishes on the boundary, and if $\mathbf{u}$ is periodic in some direction then so is $\mathbf{v}$.

Suppose also that time averages of time derivatives vanish and that time averaging commutes with spatial derivative operations. Taking the time average of (3.2.1) and (3.2.2) we find

$$\mathbf{U} \cdot \nabla \mathbf{U} + \langle \mathbf{v} \cdot \nabla \mathbf{v} \rangle + \frac{1}{\rho} \nabla P = \nu \Delta \mathbf{U} + \frac{1}{\rho} \mathbf{f}, \tag{3.2.6}$$

$$\nabla \cdot \mathbf{U} = 0, \tag{3.2.7}$$

where the mean pressure is

$$P(\mathbf{x}) = \langle p(\mathbf{x},\cdot) \rangle. \tag{3.2.8}$$

Subtracting (3.2.6) from (3.2.1) and (3.2.7) from (3.2.2) leads to the equations of motion for the fluctuations:

$$\frac{\partial \mathbf{v}}{\partial t} + \mathbf{v} \cdot \nabla \mathbf{v} + \mathbf{U} \cdot \nabla \mathbf{v} + \mathbf{v} \cdot \nabla \mathbf{U} - \langle \mathbf{v} \cdot \nabla \mathbf{v} \rangle + \frac{1}{\rho} \nabla (p - P) = \nu \Delta \mathbf{v}, \tag{3.2.9}$$

$$\nabla \cdot \mathbf{v} = 0. \tag{3.2.10}$$

The stationary Navier-Stokes equations (3.2.6) and (3.2.7) with the additional body force proportional to $\langle \mathbf{v} \cdot \nabla \mathbf{v} \rangle$ due to the turbulent fluctuations are known as *Reynolds' equations*. Noting $\mathbf{v}$'s vanishing divergence, the full stress tensor balancing the body force in (3.2.6) is

$$S_{ij}^{total} = -\delta_{ij} P + \rho\nu \left[\frac{\partial U_i}{\partial x_j} + \frac{\partial U_j}{\partial x_i} \right] - \rho \langle v_i v_j \rangle. \qquad (3.2.11)$$

Turbulence gives rise to an additional stress driving the mean flow, $-\rho\langle v_i v_j \rangle$, known as the *Reynolds stress*. In (3.2.9) and (3.2.10), we observe that the fluctuations themselves are driven by the mean flow and pressure fields, rather than being driven directly by the body force or the boundary conditions.

A hierarchy of equations relating the mean flow to various time averaged moments of the fluctuations may be derived. Indeed, in order to solve the stationary (albeit nonlinear and elliptic) problem for $\mathbf{U}$, the Reynolds stress $\langle v_i v_j \rangle$ must be supplied. A stationary equation for the Reynolds stress may be derived, but it will involve higher correlation functions such as $\langle v_i v_j v_k \rangle$. This problem continues and the hierarchy never closes because at each stage another function from a higher stage is required. This is the essence of the so-called *turbulence closure problem.* The philosophy of this approach now becomes apparent: To make any progress some closure of the hierarchy must be introduced by hand. We want to stress that any truncation of the hierarchy constitutes an *approximation* which is not derived from the original equations.

Most efforts have concentrated on developing closures at the level of the Reynolds stress because, just as a Newtonian fluid is characterized by a simple relationship between the shear stress and the rate of strain, so too it is reasonable to hope that nature strives to realize a simple relationship between the Reynolds stress and the mean velocity's strain rate. This idea is even more compelling considering that it is the mean velocity's rate of strain tensor which, on average, supplies the energy to the turbulent fluctuations. To see this, note that for either periodic or rigid boundary conditions, the turbulent kinetic energy evolves according to

$$\begin{aligned} \frac{d}{dt}\left(\frac{1}{2}\rho\|\mathbf{v}\|_2^2\right) &= -\nu\rho\|\nabla\mathbf{v}\|_2^2 - \frac{1}{2}\rho\int_\Omega v_i \left[\frac{\partial U_i}{\partial x_j} + \frac{\partial U_j}{\partial x_i}\right] v_j \, d^d x \\ &\quad -\rho\int_\Omega \frac{\partial v_i}{\partial x_j}\langle v_i v_j\rangle \, d^d x. \qquad (3.2.12) \end{aligned}$$

On average, in the steady state, the first and last terms vanish, the remaining balance being between the rate of energy supply to the turbulent fluctuations and the rate of viscous energy dissipation by the turbulent fluctuations. Viscosity dissipates turbulent kinetic energy, so there must be a significant (negative, in fact) correlation between $\langle v_i v_j \rangle$ and the mean's shear rate R_{ij} (recall (1.4.9)):

$$\int_\Omega R_{ij} \langle v_i v_j \rangle \, d^d x = -2\nu \|\nabla \mathbf{v}\|_2^2 < 0. \tag{3.2.13}$$

As an example of a closure scheme applied to a specific problem, as well as to illustrate some general features of turbulent versus laminar flows, we consider a shear-driven flow between parallel plates separated by a gap of width h. This situation is shown in Figure 3.1. A Newtonian fluid of density ρ and viscosity ν is contained between plates parallel to the $x-y$ plane, and the boundary conditions at $z=0$ and $z=h$ are

$$\mathbf{u}(x,y,0,t) = 0 \qquad \mathbf{u}(x,y,h,t) = \hat{\mathbf{i}}U. \tag{3.2.14}$$

For simplicity we do not include boundaries in the x- and y-directions, and we assume spatial homogeneity, in the mean, in these transverse directions. The Reynolds number is

$$R = \frac{Uh}{\nu}. \tag{3.2.15}$$

The solution of the problem that we seek is a mean velocity profile from which physically important quantities like the steady state average energy dissipation rate (the power expended by the agent enforcing the boundary conditions) or the wall shear stress (the force required to impose the boundary conditions) may be computed.

There is an exact solution of the Navier-Stokes equations in this geometry, known as *Couette flow*:

$$\mathbf{u} = \hat{\mathbf{i}}\frac{Uz}{h}. \tag{3.2.16}$$

This steady laminar profile is expected to be the unique time asymptotic solution at low Reynolds number.[1] The shear stress is uniform throughout Couette flow,

$$\tau = \rho\nu R_{13} = \rho\nu\frac{\partial u_1}{\partial z} = \frac{\rho\nu U}{h}. \tag{3.2.17}$$

The force required to slide the top plate over the bottom lubricated by

[1] As shown in exercise 2.5, Couette flow is nonlinearly stable at low R.

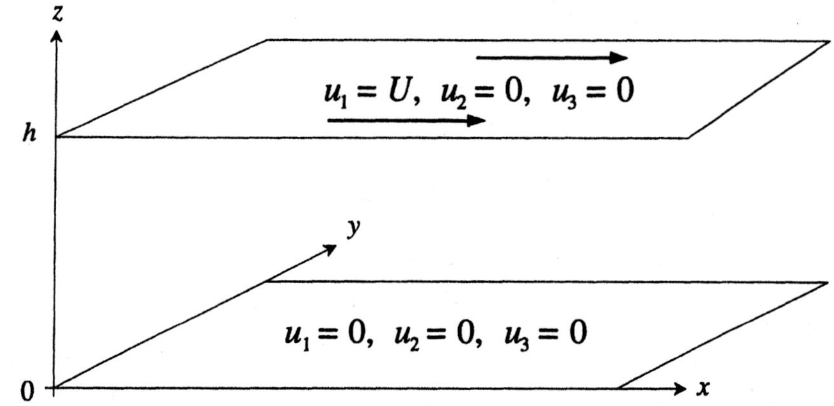

Fig. 3.1. Set-up for the analysis of turbulent shear flow between parallel plates.

the fluid, when the flow is in this laminar state, is

$$F = \tau A = \frac{\rho \nu U A}{h}, \tag{3.2.18}$$

where A is the area of the plates. This expression for the drag force is familiar from elementary physics texts, i.e., the drag is proportional to the speed, the contact surface area, and the viscosity. For later reference we put this result for the drag in a nondimensional form:

$$\tau_{nondim} = \frac{h^2 \tau}{\rho \nu^2} = \frac{Uh}{\nu} = R. \tag{3.2.19}$$

Consequently, in the laminar state at low Reynolds numbers R is precisely the nondimensional measure of the imposed stress.

At high Reynolds numbers a turbulent state is presumed to be realized, accompanied by another mean flow profile and a different Reynolds number dependence for the drag. The equations for the mean flow $\mathbf{U}(\mathbf{x})$ are then Reynolds' equations given in (3.2.6) and (3.2.7) with the boundary conditions

$$\mathbf{U}(x, y, 0) = 0, \qquad \mathbf{U}(x, y, h) = \hat{\mathbf{i}} U. \tag{3.2.20}$$

In order to proceed, some assumptions must be made.

First we impose the symmetry of the geometry on the mean flow. Translational invariance in the x- and y-directions implies that the mean flow field, the mean pressure, and the Reynolds stress are functions of

z alone. The divergence-free condition combined with the boundary conditions then immediately gives $U_3 = 0$. Isotropy in the y-direction implies $U_2 = 0$. The remaining component is $U_1(z)$ which, when translated by $U/2$, should be antisymmetric about the middle of the gap at $z = h/2$. That is, alternative boundary conditions for the mean profile between $z = 0$ and the middle of the gap at $z = h/2$ are $U_1(0) = 0$ and $U_1(h/2) = U/2$. Using these symmetry assumptions, the x component of (3.2.6) becomes

$$\frac{d}{dz}\langle v_1 v_2 \rangle = \nu \frac{d^2 U_1(z)}{dz^2}. \tag{3.2.21}$$

Integrating up from $z = 0$ and using the vanishing boundary conditions on the fluctuations, $v_1(x, y, 0, t) = v_3(x, y, 0, t) = 0$, we obtain

$$\langle v_1 v_2 \rangle(z) = \nu U_1'(z) - \nu U_1'(0), \tag{3.2.22}$$

where the prime denotes derivative with respect to z. The last term above is proportional to the mean wall shear stress,

$$\langle \tau \rangle = \rho \nu U_1'(0), \tag{3.2.23}$$

which is to be determined.

At this stage, some closure must be introduced. Along the general lines discussed earlier, we inject an ad hoc functional relationship between the Reynolds stress and the mean shear. By analogy with viscous shear stresses we assume a relation

$$\langle v_1 v_2 \rangle(z) = -\mu(z) U_1'(z), \tag{3.2.24}$$

introducing the proportionality factor $\mu(z)$, with units of viscosity $((\text{length})^2 \times (\text{time})^{-1})$, referred to as the *eddy viscosity*. The eddy viscosity should be positive: The minus sign in (3.2.24) above is a consequence of (3.2.13) which requires that $U_1'\langle v_1 v_2 \rangle$ should be negative on average.

The eddy viscosity is a property of the flow field so we should construct it out of local quantities determined by the mean flow and the geometry, and *not* from material parameters. Because the mean shear rate, $U_1'(z)$, has the dimension of inverse time, we choose this as a natural inverse time scale. A natural length scale, called the *mixing length*, is given by z, the distance to the rigid wall. The eddy viscosity is then defined

$$\mu(z) = \kappa^2 z^2 U_1'(z), \tag{3.2.25}$$

where the dimensionless constant κ is a fitting parameter in this theory, known as the *von Karman constant*, which must be determined by experiment. (It has a nominal value of $\kappa = 4/10$.) The closure that we make is thus

$$\langle v_1 v_2 \rangle(z) = -\kappa^2 z^2 \left[U_1'(z)\right]^2. \tag{3.2.26}$$

Note that this closure asssumption is to be used in the lower half of the flow where $0 \le z \le h/2$. For z between $h/2$ and h, the mixing length z should be replaced by $h - z$.

For $0 \le z \le h/2$, equation (3.2.22) then becomes

$$0 = \kappa^2 z^2 \left[U_1'(z)\right]^2 + \nu U_1'(z) - u_*^2, \tag{3.2.27}$$

where u_* is the velocity scale defined by the (as yet) unknown wall shear stress:

$$u_* = \left(\frac{\langle\tau\rangle}{\rho}\right)^{1/2} = \left[\nu U_1'(0)\right]^{1/2}. \tag{3.2.28}$$

Solving the quadratic equation in (3.2.27) for U_1' as a function of z and integrating from $z = 0$, we find the flow profile

$$U_1(z) = \frac{\kappa}{u_*} \log\left[\frac{2\kappa u_* z}{\nu} + \sqrt{1 + \left(\frac{2\kappa u_* z}{\nu}\right)^2}\right] + \frac{1 - \sqrt{1 + \left(\frac{2\kappa u_* z}{\nu}\right)^2}}{\frac{2\kappa u_* z}{\nu}}. \tag{3.2.29}$$

The velocity scale u_* is fixed by the boundary condition $U_1(h/2) = U/2$:

$$\frac{1}{2}\kappa\left(\frac{u_*}{U}\right)^{-1} = \log\left[\kappa R \frac{u_*}{U} + \sqrt{1 + \left(\kappa R \frac{u_*}{U}\right)^2}\right] + \frac{1 - \sqrt{1 + \left(\kappa R \frac{u_*}{U}\right)^2}}{\kappa R \frac{u_*}{U}}. \tag{3.2.30}$$

For a given Reynolds number, (3.2.30) is to be solved for $\frac{u_*}{U}$. Then (3.2.29) and (3.2.28) yield explicit predictions for the mean flow profile and the wall shear stress. A typical profile at moderately high Reynolds number is shown in Figure 3.2. The shear in the mean flow is concentrated in thin layers near the rigid boundaries. At large Reynolds numbers, these boundary layers have thickness $\delta \sim \frac{\nu}{u_*}$. The turbulence intensity, as measured by the correlation $\langle v_1 v_2 \rangle$ in (3.2.26), is sketched in Figure 3.3. Turbulent fluctuations are inhibited by the presence of the walls, and so are greatest near the center of the gap. Note that the correlations in the turbulent fluctuations are on the scale of u_*.

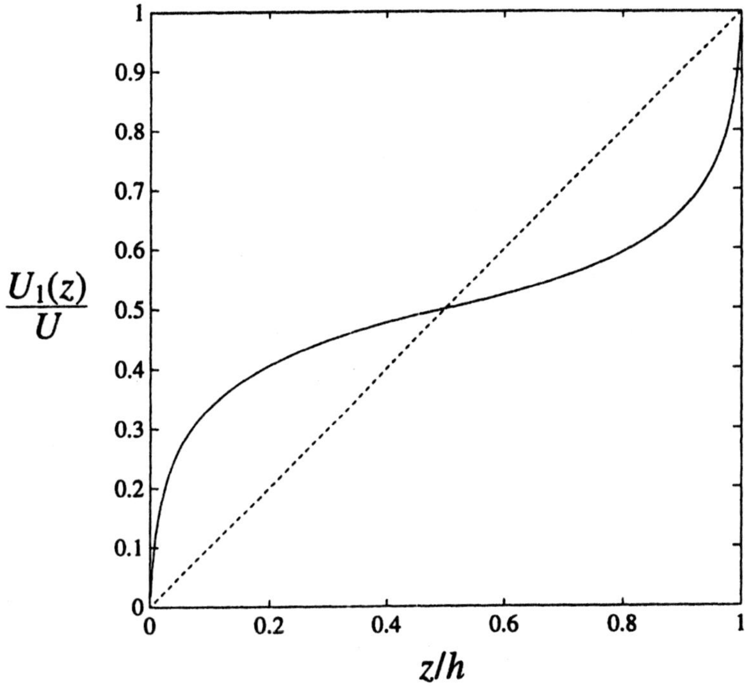

Fig. 3.2. Mean flow profiles from the closure approximation (solid line) for $R = 10^4$. The dashed line is the laminar Couette flow profile.

In the limit $R \to 0$, the solution of (3.2.30) is

$$u_* \to \frac{U}{R^{1/2}}, \tag{3.2.31}$$

yielding precisely the laminar stress in (3.2.17) and (3.2.19). The fitting parameter κ does not enter into this limit of the model. In the opposite limit, $R \to \infty$, (3.2.30) yields

$$u_* \to \kappa \frac{U}{\log R}, \tag{3.2.32}$$

and the nondimensional drag force in this limit is

$$\frac{h^2 \tau}{\rho \nu^2} \to \kappa^2 \frac{R^2}{(\log R)^2}. \tag{3.2.33}$$

Asymptotically at high Reynolds numbers, this theory predicts a turbulent drag force proportional to the square of the speed U with logarithmic

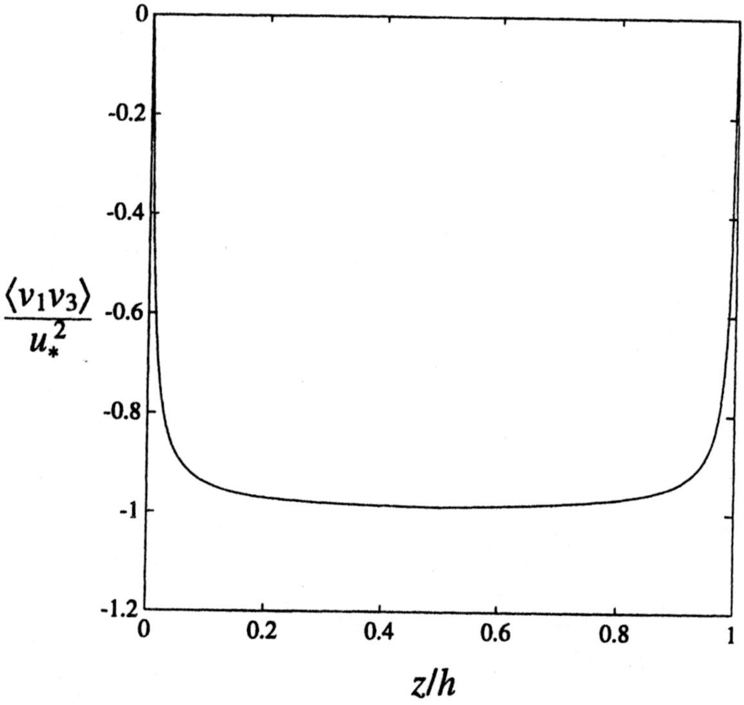

Fig. 3.3. Reynolds stress across the layer. Note that u_* is the scale of the velocity fluctuations.

corrections. The drag force predictions[1] are plotted over several orders of magnitude of Reynolds number in Figure 3.4.

Although this example has been developed for illustrative purposes with a minimum of fitting parameters (one), it does reproduce some essential features of the conventional wisdom concerning the structure of both the mean and Reynolds stress profiles. Together with logarithmic corrections, the R^2 scaling of the turbulent drag is supported by experiments in wall bounded shear flows. The introduction of an eddy viscosity, a mixing length, and a closure as in (3.2.26), however, constitute uncontrolled approximations which do not follow rigorously (or even "formally" or systematically) from the Navier-Stokes equations. A central challenge for Navier-Stokes analysis is to derive these kinds of results and predictions directly from the equations of motion without making such additional assumptions.

[1] Comparison with experimental data is made in Chapter 10.

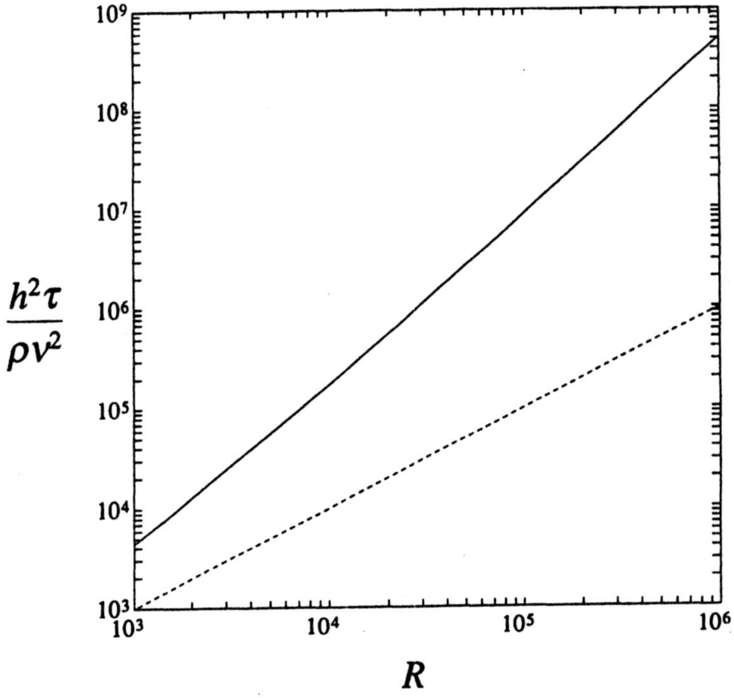

Fig. 3.4. Drag versus Reynolds number from the closure approximation (solid line). The drag from the laminar Couette flow is shown for comparison (dashed line).

3.3 Spectra, Kolmogorov's scaling theory, and turbulent length scales

Another common tool in turbulence theory is spectral analysis. Here we consider the translational invariant case of periodic boundary conditions on $\Omega = [0, L]^d$. The Fourier transform of the velocity vector field is

$$\hat{\mathbf{u}}(\mathbf{k}, t) = \int_\Omega e^{-i\mathbf{k}\cdot\mathbf{x}} \mathbf{u}(\mathbf{x}, t)\, d^d x, \tag{3.3.1}$$

with inverse transform

$$\mathbf{u}(\mathbf{x}, t) = L^{-d} \sum_{\mathbf{k}} e^{i\mathbf{k}\cdot\mathbf{x}} \hat{\mathbf{u}}(\mathbf{k}, t), \tag{3.3.2}$$

where the discrete wavenumbers $\mathbf{k}$ are, in $3d$,

$$\mathbf{k} = \hat{\mathbf{i}}\frac{2\pi n_1}{L} + \hat{\mathbf{j}}\frac{2\pi n_2}{L} + \hat{\mathbf{k}}\frac{2\pi n_3}{L} \tag{3.3.3}$$

with $n_i = 0, \pm 1, \pm 2, \ldots$. The Fourier coefficients $\hat{\mathbf{u}}$ are the complex amplitudes of the *modes* labeled by the wavenumber $\mathbf{k}$.

Rather than considering the fluid motion in physical space, its representation in wavenumber space is a decomposition of the flow field into components at various length scales, the length scale associated with wavenumber $\mathbf{k}$ being $2\pi/|\mathbf{k}|$. Thus high wavenumbers correspond to small spatial scales, whereas small wavenumbers correspond to large scales. For example, using Parseval's equality, the instantaneous kinetic energy is

$$\frac{1}{2}\rho\|\mathbf{u}(\cdot,t)\|_2^2 = \frac{1}{2}\rho L^{-d}\sum_{\mathbf{k}}|\hat{\mathbf{u}}(\mathbf{k},t)|^2. \tag{3.3.4}$$

The total kinetic energy is the sum of the energies in the modes. In the limit of large volumes where the volume elements in wavenumber space become infinitesimal, and the sum over wavenumbers can be converted to an integral, the total energy is (note that the volume element in wavenumber space is $d^dk = [2\pi/L]^d$)

$$\frac{1}{2}\rho\|\mathbf{u}(\cdot,t)\|_2^2 = \frac{1}{2}\int_{k\geq 2\pi/L}\frac{\rho}{(2\pi)^d}|\hat{\mathbf{u}}(\mathbf{k},t)|^2 d^dk. \tag{3.3.5}$$

Because the quantity $\frac{1}{2}\rho|\mathbf{u}(\mathbf{x},t)|^2$ is naturally recognized as the local kinetic energy density in real space, (3.3.5) leads to the identification of

$$\frac{1}{2}\frac{\rho}{(2\pi)^d}|\hat{\mathbf{u}}(\mathbf{k},t)|^2 \tag{3.3.6}$$

as the kinetic energy density in wavenumber space. Integrated over a thin shell around $|\mathbf{k}| = k$ in wavenumber space, this density gives the contribution to the kinetic energy in the flow field from motion on the length scale $2\pi/k$.

For the moment let us focus on the 3d case. If the turbulence is isotropic,[1] or is at least approximately so at small length scales, then the average local energy in those modes is only a function of the wavenumber amplitude k, and we may define the wavenumber spectrum for the energy per unit mass by

$$\begin{aligned}\frac{1}{2L^3}\int\frac{1}{(2\pi)^3}\langle|\hat{\mathbf{u}}(\mathbf{k},\cdot)|^2\rangle\, d^3k &= \frac{1}{2L^3}\int\frac{1}{(2\pi)^3}\langle|\hat{\mathbf{u}}(\mathbf{k},\cdot)|^2\rangle(4\pi k^2\, dk)\\ &=: \int E(k)dk.\end{aligned} \tag{3.3.7}$$

[1] Isotropic turbulence means that on average, there is no preferred direction in the flow.

The density $E(k)$ is the contribution to the total average kinetic energy per unit mass from each length scale $2\pi/k$ and the total is

$$\frac{1}{2L^3}\langle\|\mathbf{u}\|_2^2\rangle = \int_{k\geq 2\pi/L} E(k)\,dk. \tag{3.3.8}$$

The time evolution of the different wavenumber components is computed as the Fourier transform of the Navier-Stokes equations (3.2.1) and (3.2.2)

$$\frac{d}{dt}\hat{\mathbf{u}}(\mathbf{k},t) = -\nu k^2\hat{\mathbf{u}}(\mathbf{k},t) + \frac{i}{L^3}\left(\mathbf{I}-\frac{\mathbf{kk}}{k^2}\right)\cdot\sum_{\mathbf{k}'+\mathbf{k}''=\mathbf{k}}\left[\hat{\mathbf{u}}(\mathbf{k}',t)\cdot\mathbf{k}''\hat{\mathbf{u}}(\mathbf{k}'',t)\right] + \hat{\mathbf{f}}(\mathbf{k}), \tag{3.3.9}$$

$$\mathbf{k}\cdot\hat{\mathbf{u}}(\mathbf{k},t) = 0, \tag{3.3.10}$$

where $\mathbf{I}$ is the unit tensor, $\mathbf{I}-\mathbf{kk}/k^2$ is the projector onto divergence free vector fields in wavenumber space, and $\hat{\mathbf{f}}(\mathbf{k})$ is the Fourier transform of the body force (without loss of generality, $\mathbf{k}\cdot\hat{\mathbf{f}}(\mathbf{k}) = 0$).[1] This formulation leads to a useful interpretation of the dynamics in the Navier-Stokes equations because the evolution of the flow on various length scales can be separated into three distinct terms each playing a different role.

First, the body force provides a direct local stimulation to the modes at wavenumbers in the support of $\hat{\mathbf{f}}(\mathbf{k})$. This straightforward mechanism provides the energy which will be continuously consumed by the turbulence and which is essential to its maintenance. For the purposes of the discussion here, we will consider the influence of the body force to be significant only on relatively long length scales where $kL \sim O(1)$, its sole purpose being to supply the power to drive the system. The evolution of the amplitudes for shorter scales is then determined by the other terms in the Navier-Stokes equations.

The Fourier transform of the nonlinear terms ($\mathbf{u}\cdot\nabla\mathbf{u}$ and the pressure gradient) leads to mode coupling in wavenumber space. This coupling provides a mechanism for the participation of higher wavenumber modes not directly excited by the body force. In real space, the nonlinear terms instigate instabilities of long wavelength, low modal flow configurations which may lead to perturbations at some shorter scales. As the shorter scales are excited, the "local" Reynolds number may again exceed some critical value, leading the instability to even shorter wavelengths: That is, the nonlinear terms can lead to a transfer of energy from long to

[1] Note that the Fourier transform of the pressure does not appear explicitly in (3.3.9). Rather, it contributes to the projection operator. The origin of the pressure in terms of the incompressibility constraint is readily apparent in the Fourier transformed formulation.

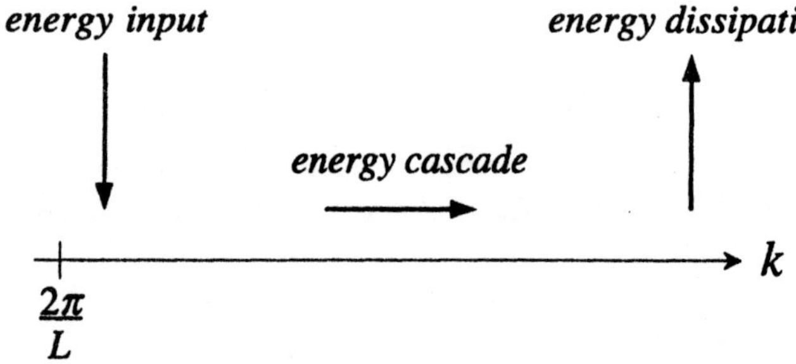

Fig. 3.5. Schematic illustration of the cascade process in $3d$ turbulence.

short length scales. This nonlinear energy transfer between the modes is conservative, and no energy is consumed in these interactions.

Another view of the nonlinear process of energy transfer between various scales comes from looking at the interplay of energy conservation and vortex stretching in the underlying inviscid system. Consider a "packet" of energy initially concentrated at relatively low k values and neglect both the influence of the body force at lower k's and the effect of viscosity at very high k's. The total energy $\int E(k)dk$ is conserved in the subsequent evolution, while vortex stretching works to increase the enstrophy

$$\|\omega\|_2^2 \sim \Omega \int k^2 E(k)\, dk. \tag{3.3.11}$$

The only way that $\int E(k)dk$ can remain constant while $\int k^2 E(k)dk$ increases is for more and more of the energy to be found at higher and higher wavenumbers. This *cascade* process is illustrated schematically in Figure 3.5.

The viscous dissipation term, $-\nu k^2 \hat{\mathbf{u}}(\mathbf{k}, t)$, damps the excitations, though with very different influence at different length scales. On relatively long length scales, which go out to the "integral" or "outer" scale L, the viscous term has much less damping effect than on short scales. Indeed, turbulent flows are characterized by high Reynolds numbers and some simple considerations give an indication of the key role played by a Reynolds number. Denoting a typical fluctuation velocity scale of the flow by U, (3.3.9) implies that the influence of the viscosity is negligible compared to the nonlinear term at wavenumbers where

$$\nu k^2 \ll Uk, \tag{3.3.12}$$

which is equivalent to

$$kL \ll R \tag{3.3.13}$$

for the Reynolds number based on U, L and ν. On the other hand, viscosity dominates when

$$\nu k^2 \gg Uk, \tag{3.3.14}$$

i.e., at short wavelengths where

$$kL \gg R. \tag{3.3.15}$$

The shorter the length scale, the higher the k, and the stronger the damping effect of viscosity. Thus a Reynolds number roughly defines the border between the wavenumbers of modes whose evolution is dominated by either viscosity or by nonlinearities. We caution the reader not to take the length scale boundaries in (3.3.12)-(3.3.15) too literally at this stage, for the velocity scale used in this Reynolds number is only vaguely defined. A more sophisticated theory, outlined below, gives a more precise picture of the length and velocity scales associated with turbulent flows.

The cascade of energy from long to short wavelengths, a salient feature of turbulent dynamics, is believed to lead to some universal features. A fundamental theory of ubiquitous structure in homogeneous isotropic turbulence is *Kolmogorov's scaling theory* which we paraphrase as follows. In a steady state, energy provided by the body force at small wavenumbers is flowing through the system from small to large k's, to be dissipated by viscosity at high wavenumbers. The energy input rate, which on average in the steady state is the same as the energy dissipation rate, is the key external quantity determining the intensity of the turbulence. We use ε to denote the average energy dissipation rate per unit mass

$$\varepsilon = L^{-3}\nu\langle \|\nabla \mathbf{u}\|_2^2\rangle, \tag{3.3.16}$$

with units of $(\text{length})^2 \times (\text{time})^{-3}$. We assume that at length scales shorter than those directly excited, but not so short as to be viscous dominated, the local structure of the energy density spectrum in wavenumber space, $E(k)$, is independent of the viscous dissipation mechanism. On these intermediate scales, in the so-called *inertial range*, the structure of the energy density is then determined solely by the nonlinear evolution and the overall energy flux through the system. This means that $E(k)$ must be constructed out of ε and k alone with no explicit reference to the viscosity ν or the outer length L. Recalling (3.3.8), $E(k)$ has units

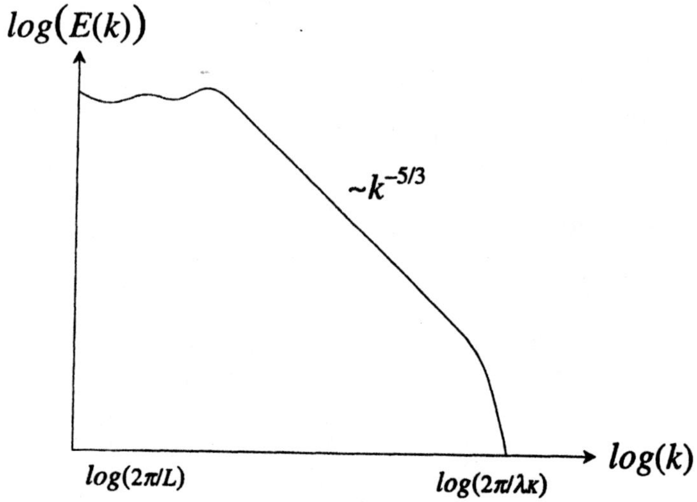

Fig. 3.6. Schematic structure of the energy spectrum in Kolmogorov's scaling theory.

$(\text{length})^3 \times (\text{time})^{-2}$, and thus dimensional analysis inplies that the energy density must be

$$E(k) = C_K \varepsilon^{2/3} k^{-5/3}, \tag{3.3.17}$$

where C_K is a "universal" dimensionless constant.

The inertial range, defined by a $k^{-5/3}$ form of the spectrum, should extend down to a length scale where viscous energy consumption effectively cuts off the energy (or where it at least induces a much faster, presumably exponential, decay of $E(k)$). This cut-off scale must depend primarily on ε and ν, becoming a smaller length with either increasing ε or decreasing ν. The unique length scale that can be constructed out of ε and ν is called the *Kolmogorov dissipation length*:

$$\lambda_K = c_K \left(\frac{\nu^3}{\varepsilon} \right)^{1/4}, \tag{3.3.18}$$

where c_K is another dimensionless constant. The energy spectrum in the Kolmogorov scaling theory is shown in Figure 3.6. It consists of some nonuniversal structure at low wavenumbers $\sim 2\pi/L$, which depend on the details of the forcing, an inertial range characterized by the $k^{-5/3}$ scaling of $E(k)$, and a viscous cut-off of energy at $2\pi/\lambda_K$.

The factors C_K and c_K in (3.3.17) and (3.3.18) are not independent. A

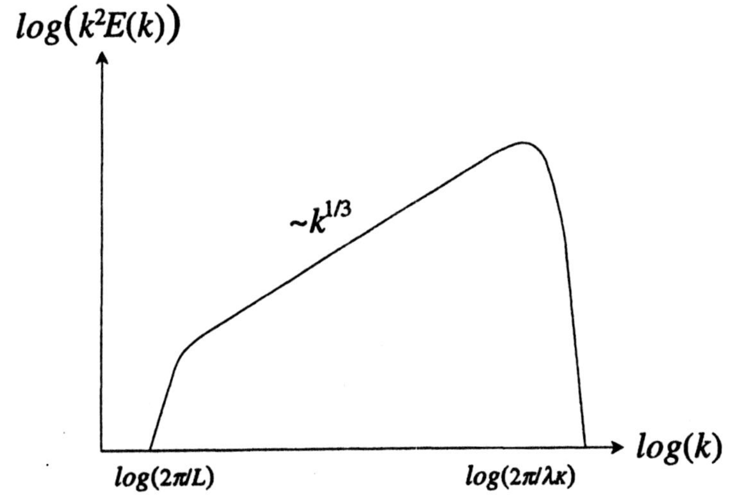

Fig. 3.7. Structure of the enstrophy spectrum, which is the energy dissipation rate spectrum, in Kolmogorov's theory.

simple relationship between them follows from the observation that the energy dissipation rate is self-consistently related to the energy spectrum according to

$$\varepsilon = 2\nu \int_{k \geq 2\pi/L} E(k)k^2\, dk. \tag{3.3.19}$$

The associated energy dissipation wavenumber spectrum, $2\nu k^2 E(k)$, is illustrated in Figure 3.7. The dissipation takes place predominantly in the inertial range and is also essentially cut off at the Kolmogorov scale, so we have

$$\varepsilon \approx 2C_K \varepsilon^{2/3} \nu \int_0^{2\pi/\lambda_K} k^{1/3}\, dk = \frac{3}{2} \left(\frac{2\pi}{\lambda_K} \right)^{4/3} C_K \varepsilon^{2/3} \nu. \tag{3.3.20}$$

Inserting (3.3.18), we conclude that

$$c_K = 2\pi \left(\frac{3}{2} C_K \right)^{3/4}. \tag{3.3.21}$$

Hence, Kolmogorov's scaling theory for the spectrum of homogeneous turbulence has essentially one free parameter, to be determined by other methods or by a fit to experimental data.

The Kolmogorov spectrum also allows for the definition of a Reynolds number by providing a velocity scale from the system parameters ε, ν

and L. Indeed, the total kinetic energy per unit mass defines a velocity scale U according to

$$\frac{1}{2}U^2 = \int_{2\pi/L}^{2\pi/\lambda_K} E(k)\,dk \sim \varepsilon^{2/3}L^{2/3}. \tag{3.3.22}$$

Hence the velocity scale is

$$U \sim (\varepsilon L)^{1/3}, \tag{3.3.23}$$

and the Reynolds number, based on U, L and ν, is

$$R = \frac{\varepsilon^{1/3}L^{4/3}}{\nu}. \tag{3.3.24}$$

The Kolmogorov length may then be expressed very simply in terms of the outer length L and the Reynolds number:

$$\lambda_K \sim LR^{-3/4}. \tag{3.3.25}$$

Thus, referring back to the discussion around (3.3.12) to (3.3.15), we see that Kolmogorov's theory predicts that viscous dominated evolution sets in with modes of wavenumber k_c satisfying

$$k_c L \sim R^{3/4}. \tag{3.3.26}$$

This is a more precise result than that following from the discussion around (3.3.12) to (3.3.15): The source of the discrepancy between (3.3.26) and (3.3.12)-(3.3.15) is the choice of velocity scales entering the definition of the Reynolds number.[1]

Another small length scale defined by the turbulent energy spectrum is given by the mean square width of the energy spectrum. This scale is known as the *Taylor micro-scale* λ_T:

$$\lambda_T^{-2} \sim \frac{\int k^2 E(k)dk}{\int E(k)dk} = \frac{\langle \|\nabla \mathbf{u}\|_2^2\rangle}{\langle \|\mathbf{u}\|_2^2\rangle} = \frac{\langle \|\boldsymbol{\omega}\|_2^2\rangle}{\langle \|\mathbf{u}\|_2^2\rangle}. \tag{3.3.27}$$

Within Kolmogorov's theory, using $\varepsilon \sim U^3/L$ from (3.3.23)

$$\lambda_T \sim \left(\frac{U^2}{U^3/\nu L}\right)^{1/2} = LR^{-1/2}. \tag{3.3.28}$$

This scale is larger than the dissipation length scale, and is more indicative

[1] For the situation described here, R is not directly controlled externally. Rather, the body force (measured by the Grashof number) is imposed and the flow organizes itself into a state with Reynolds number given by (3.3.24). This theory does not directly predict the Reynolds number as a function of the Grashof number.

of the lengths above which a significant amount of kinetic energy is concentrated.

The Kolmogorov length, $\sim R^{-3/4}$, and the Taylor micro-scale, $\sim R^{-1/2}$, may not be the only small scales in turbulent flow: They are identified here within a theory for homogeneous isotropic turbulence. If there are rigid boundaries, then boundary layers may play an important role in turbulent dynamics. For example, in the approximate theory of shear driven turbulence developed in the previous section, the boundary layer thickness is $\sim R^{-1}$ (to within logarithms) in terms of the Reynolds number based on the outer length and velocity scales. This is an even smaller scale than the Kolmogorov length.

For all these problems, typical challenges for Navier-Stokes analysis are

- To give a more systematic derivation of Kolmogorov's picture from the Navier-Stokes equations, and
- For specific systems, to give predictions for ε, or length scales such as λ_K, λ_T or boundary layer thicknesses, in terms of externally controlled stress parameters.

The situation is very different for the $2d$ Navier-Stokes equations due to the absence of the vortex stretching mechanism. In an unforced inviscid $2d$ flow both the energy and the enstrophy are conserved, so we can no longer appeal to an energy cascade in wavenumber space to compel us to adopt the energy flux through the system as the key physical parameter. Indeed, if both $\int E(k)dk$ and $\int k^2 E(k)dk$ are constants of the motion, then any flow of energy from low to high wavenumbers must be accompanied by another compensating flux of energy back toward *larger* length scales. This feature of $2d$ turbulence is referred to as the *inverse cascade*.

Although the $2d$ bulk dynamics does not conspire to enhance the enstrophy, it does have a mechanism for amplifying *gradients* of vorticity.[1] Referring to the inviscid evolution (1.3.16), the gradients of ω evolve according to

$$\left(\frac{\partial}{\partial t} + \mathbf{u} \cdot \nabla\right) \frac{\partial \omega}{\partial x_i} = -\frac{\partial u_j}{\partial x_i} \frac{\partial \omega}{\partial x_j}. \tag{3.3.29}$$

In analogy with vortex stretching, we see that $\nabla\omega$ can increase when it is properly aligned with the rate of strain tensor. That is, the L^2 norm of

[1] Enstrophy in $2d$ flows can be generated at rigid boundaries.

the gradient of the vorticity evolves according to

$$\frac{d}{dt}\left(\frac{1}{2}\|\nabla\omega\|_2^2\right) = -\int \frac{\partial\omega}{\partial x_i} R_{ij} \frac{\partial\omega}{\partial x_j}\, d^2x, \tag{3.3.30}$$

where R_{ij} is the flow field's symmetric rate of strain tensor. Thus $\|\nabla\omega\|_2^2$ may spontaneously increase in time. The L^2 norm of the gradient of the vorticity is related to $\int k^4 E(k)dk$ and, in the viscous case, to the *enstrophy* dissipation rate

$$\frac{d}{dt}\left(\frac{1}{2}\|\omega\|_2^2\right) = -\nu\|\nabla\omega\|_2^2. \tag{3.3.31}$$

Compared to the $3d$ case, we see that the enstrophy in $2d$ plays a role previously played by the energy: Its flux through the system, or its dissipation rate, is the quantity which the rate of strain tensor can amplify in a cascade process. By analogy, then, we may formulate a high wavenumber dissipation scaling theory for $2d$ turbulence based on the average enstrophy dissipation rate defined by

$$\chi = \nu L^{-2}\langle\|\nabla\omega\|_2^2\rangle. \tag{3.3.32}$$

The idea is that the high wavenumber universal aspects of the energy spectrum are determined solely by the enstrophy flux and the local wavenumber and, at the very highest wavenumbers, by the the enstrophy flux and the viscosity. The rest follows from dimensional analysis. Because χ has units of $(\text{time})^{-3}$ and $E(k)$ has units $(\text{length})^3 \times (\text{time})^{-2}$, the high wavenumber energy spectrum must be

$$E(k) \sim \chi^{2/3}k^{-3}. \tag{3.3.33}$$

The dissipation length scale, below which viscosity dominates the dynamics, is the unique scale determined by χ and ν, known as the *Kraichnan length*

$$\lambda_{Kr} \sim \left(\frac{\nu^3}{\chi}\right)^{1/6}. \tag{3.3.34}$$

The low wavenumber inverse cascade implied by the inviscid conservation of enstrophy in $2d$ obeys a different law. In a steady state situation one imagines a body force providing energy and enstrophy for the turbulence at a length scale intermediate to the outer scale L and an "inner" scale like λ_{Kr}. Enstrophy flows toward high wavenumbers and its flux establishes the high wavenumber scaling. Energy flows toward longer scales and its flux determines the low wavenumber scaling. By dimensional analysis then, the unique $E(k)$ fixed by ε and k is the

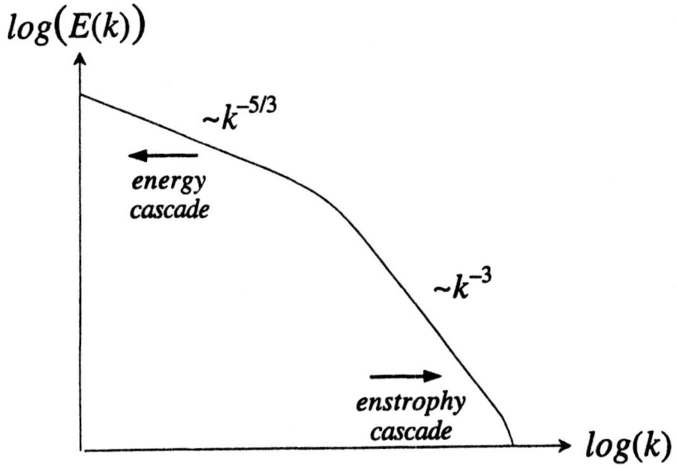

Fig. 3.8. The energy spectrum in $2d$ turbulence.

Kolmogorov spectrum

$$E(k) \sim \varepsilon^{2/3} k^{-5/3}. \tag{3.3.35}$$

This structure for the energy spectrum is sketched in Figure 3.8. We caution the reader, however, that this picture is not really self-consistent as far as the existence of a steady state is concerned. This is because a constant flux of energy to low k implies a continual piling up of energy in the lowest modes, with no mechanism for dissipating it. Thus no real steady state can be achieved within the enstrophy-cascade/inverse-energy-cascade scenario. This problem is characteristic of the open questions that remain for the development of a general scaling theory of $2d$ turbulence. In any case, both χ and ε are quantities essential to the characterization of $2d$ turbulence. Future problems for the rigorous analysis of $2d$ turbulence include systematically estimating χ, ε and λ_{Kr} as functions of some externally controlled Grashof, Reynolds, or Rayleigh number which appropriately measure the applied stress.

3.4 References and further reading

Conventional $3d$ turbulence theories are discussed, for example, in Tennekes and Lumley [7]. Two-dimensional turbulence is discussed in Kraichnan and Montgomery [8].

Exercises

1 Verify (3.2.31) and (3.2.32) and the corresponding limits of the drag as a function of R for the closure approximation developed in section 3.2.

2 Consider the stationary flow field

$$\mathbf{U}(\mathbf{x}) = \hat{\mathbf{i}}\frac{Uz(h-z)}{h^2} \tag{E3.1}$$

in the region $0 \leq z \leq h$. Show that this is an exact stationary solution of the Navier-Stokes equations for a fluid confined between stationary plates at $z = 0$ and $z = h$, and subject to a pressure gradient in the x-direction. Compute the "friction coefficient"

$$C_f := \frac{\Delta P/L}{\overline{U}^2/h}, \tag{E3.2}$$

where $\Delta P/L$ is the imposed gradient, and $\overline{U}$ is the mean velocity as a function of the Reynolds number

$$R = \frac{v\overline{U}}{h}. \tag{E3.3}$$

3 Starting from the definitions of the Fourier transform and its inverse in (3.3.1) and (3.3.2), derive the transformed Navier-Stokes equations (3.3.9) and (3.3.10).

4 Show that the quadratic term in $\hat{\mathbf{u}}$ in (3.3.9) neither creates nor destroys kinetic energy.

4

Degrees of freedom, dynamical systems, and attractors

4.1 Introduction

The classical Kolmogorov picture asserts that high Reynolds number $3d$ turbulence generates structures, "eddies," down to the size of the cut-off length $\lambda_K \sim LR^{-3/4}$. Below this scale viscosity effectively consumes all the energy flowing in from larger scales. A complete specification of the state of a turbulent velocity vector field must then resolve excitations on lengths at least down to this scale. This means that in a box of volume L^3, at least

$$N = \left(\frac{L}{\lambda_K}\right)^3 \sim R^{9/4} \tag{4.1.1}$$

coordinates are necessary to accurately characterize the instantaneous state of the flow. This is the origin of the claim that $3d$ turbulence has "$\sim R^{9/4}$ degrees of freedom." In $2d$ turbulence the number of degrees in a box of size $L \times L$ is taken to be the "number of eddies" of diameter Kraichnan's length, λ_{K_r}, that fit into the box:

$$N = \left(\frac{L}{\lambda_{K_r}}\right)^2. \tag{4.1.2}$$

These measures of the number of degrees of freedom in turbulent dynamics, or something closely related to them, are of interest and importance for a number of reasons. From the point of view of experiment and application they are a measure of the quantity of information or of the number of measurements required for a complete description of a flow field. In numerical analysis these quantities are a measure of the computational resolution, either in terms of grid points or Fourier modes, necessary for reliable simulations. From a more theoretical point of view, modern dynamical systems theory is generally concerned with low dimensional models corresponding to relatively few degrees of freedom. Even if turbulence is not characterized by very low dimensional models, it is certainly natural to wonder how far this theory has to go to tackle the full problem. It is thus of fundamental scientific interest to ask if fluid turbulence is truly in the domain of application of finite dimensional dynamical systems theory, and to apply what tools we can to problems in fluid mechanics.

In this chapter we present a practical definition of the notion of the number of degrees of freedom for a dynamical system – the attractor dimension – and show how it may be analytically estimated. These tools are then applied to the problem of thermal convection in the form of the Lorenz equations to illustrate the techniques on a familiar model. The natural extension of these methods to PDEs can be found in Chapter 9 where we estimate the attractor dimension of the $2d$ and $3d$ Navier-Stokes equations.

4.2 Dynamical systems, attractors, and their dimension

The standard definition of the number of degrees of freedom for Hamiltonian systems is the number of pairs of conjugate dynamical variables in the model. For instance a one Hamiltonian degree of freedom system is one particle in one dimension with Hamiltonian $H(p,q)$ and evolution defined by Hamilton's equations,

$$
\begin{aligned}
\frac{dq}{dt} &= \frac{\partial H(p,q)}{\partial p}, \\
\frac{dp}{dt} &= -\frac{\partial H(p,q)}{\partial q}.
\end{aligned}
\tag{4.2.1}
$$

In general n particles in d spatial dimensions constitute a $d \times n$ Hamiltonian degree of freedom system defined by $2dn$ first order ordinary differential equations (ODEs), two for each pair $(q_i, p_i), i = 1, \ldots, n$. The system's *phase space* is spanned by all $2dn$ of the generalized coordinates and momenta. Of course constraints and conservation laws effectively decrease the numbers of degrees of freedom realized by any particular solution. The net number of degrees of freedom is then naturally identified as the dimension of the set in phase space in which individual trajectories lie.

For more general continuous time dynamical systems the number of degrees of freedom is often associated with the number of first order ODEs in the model, net constraints. For example, an autonomous n-

dimensional dynamical system may be defined by the ODEs

$$\begin{aligned} \frac{dx_1}{dt} &= f_1(x_1,\ldots,x_n) \\ &\vdots \\ \frac{dx_n}{dt} &= f_n(x_1,\ldots,x_n). \end{aligned} \tag{4.2.2}$$

If the vector field of the velocities f_i is sufficiently smooth (e.g., Lipschitz continuous) the system evolves uniquely from n given initial conditions $x_1(0),\ldots,x_n(0)$ in an n-dimensional phase space and the state of the system at time t is specified precisely by the n coordinates $x_1(t),\ldots,x_n(t)$. It is thus natural to refer to this as a system of n degrees of freedom.

One of the most elementary classifications of dynamical systems is according to whether or not the associated evolution preserves volume in the system phase space. Recalling the discussion around (1.2.7) in Chapter 1, which is easily generalized to an arbitrary number of variables, we see that the criteria for this are very simple:

$$\frac{\partial f_i}{\partial x_i} < 0 \quad \Leftrightarrow \quad n\text{-dimensional volumes contract.} \tag{4.2.3}$$

$$\frac{\partial f_i}{\partial x_i} = 0 \quad \Leftrightarrow \quad n\text{-dimensional volumes are preserved.} \tag{4.2.4}$$

$$\frac{\partial f_i}{\partial x_i} > 0 \quad \Leftrightarrow \quad n\text{-dimensional volumes expand.} \tag{4.2.5}$$

The divergence of the velocity vector field in phase space is thus a crucial quantity. Hamiltonian dynamics are examples of volume preserving systems because the divergence of the phase space velocity vector field for the system in (4.2.1) vanishes identically.

In many physical systems the effect of friction or other dissipative agents is to cause a contraction of phase space volumes during the evolution. To illustrate this we may add a frictional influence to the Hamiltonian system in (4.2.1) by including a force, proportional in magnitude but opposite in direction to the velocity, in the rate of change of momentum equation:

$$\frac{dp}{dt} \to -\frac{\partial H(p,q)}{\partial q} - \alpha\frac{\partial H(p,q)}{\partial p}, \tag{4.2.6}$$

where the friction coefficient α is positive. Then the divergence of the phase space velocity vector field is

$$\frac{\partial}{\partial q}\left(\frac{\partial H(p,q)}{\partial p}\right)+\frac{\partial}{\partial p}\left(-\frac{\partial H(p,q)}{\partial q}-\alpha\frac{\partial H(p,q)}{\partial p}\right)=-\alpha\frac{\partial^2 H(p,q)}{\partial p^2}. \quad (4.2.7)$$

For Hamiltonians quadratic in the momenta and bounded from below (e.g., $H = p^2/2m + V(q)$ with mass $m > 0$) this is manifestly negative, so 2-volumes in phase space contract. The value of the Hamiltonian function, the energy in the system, necessarily decreases during this kind of motion,

$$\frac{dH}{dt}=\frac{\partial H(p,q)}{\partial q}\frac{dq}{dt}+\frac{\partial H(p,q)}{\partial p}\frac{dp}{dt}=-\alpha\left(\frac{\partial H(p,q)}{\partial p}\right)^2<0, \quad (4.2.8)$$

and we may expect the evolution to drive the system into a local minimum of the Hamiltonian function. Local minima lie generically on a set of dimension zero (fixed points), so initial two-dimensional sets of phase space coordinates end up on a set of lower dimension in the steady state after transients have decayed. This is not the final fate of every perturbed Hamiltonian system because it is also possible to impose driving forces which may conspire to cause (some) volume elements to expand.

An n-dimensional dynamical system with bounded trajectories in which all n-dimensional volume elements contract to zero (a sufficient condition for which is that the divergence of the phase space velocity vector field is always negative) is a *dissipative* system. Although it may seem wise to associate the "number of degrees of freedom" with the largest conceivable number of data required to configure all possible motions starting from all initial states, this is not always the most useful definition. Rather, it is natural and practical to neglect transient excitations and to focus on time asymptotic motion. For dissipative systems this asymptotic motion takes place on a set of measure zero (volume zero) in the system's configuration space, although this does not necessarily mean that the motion contracts to fixed points. It merely means that the steady state motion of a dissipative dynamical system, if such a steady state exists, is necessarily restricted to a set of lower dimension than that of the entire phase space. We refer to the set upon which the asymptotic motion takes place as the *attractor*. Roughly speaking, all bounded trajectories eventually end up either on or approaching the attractor, justifying its name. It is then natural to identify the dimension of the attractor as an effective measure of the number of degrees of freedom in the system: This definition utilizes not just the a priori "size" of the dynamical system, but also the effect of the dynamics.

Hence a task for the analysis of fluid flows, including turbulent dynamics, is to quantify the concept of an *attractor dimension*, show how it may be computed or estimated, and compare these predictions with those of conventional theories and approximations.

Note that there is another issue that must be addressed for continuum mechanical systems described by partial differential equations (PDEs); the naive count of the number of degrees of freedom based on the phase space dimension is infinite. To see this, recall the evolution equations (3.3.9)-(3.3.10) for the Fourier amplitudes of the velocity field. The PDE – in this case the Navier-Stokes equations on a periodic domain – is equivalent to the infinite set of coupled ODEs (recall (3.3.9)) for the mode amplitudes:

$$\frac{d\hat{\mathbf{u}}(\mathbf{k},t)}{dt} = -\nu k^2 \hat{\mathbf{u}}(\mathbf{k},t) + \frac{i}{L^3}\left(\mathbf{I} - \frac{\mathbf{k}\mathbf{k}}{k^2}\right) \cdot \sum_{\mathbf{k}'+\mathbf{k}''=\mathbf{k}} \hat{\mathbf{u}}(\mathbf{k}',t)\cdot \mathbf{k}''\hat{\mathbf{u}}(\mathbf{k}'',t) + \hat{\mathbf{f}}(\mathbf{k}), \tag{4.2.9}$$

$$\mathbf{k}\cdot\hat{\mathbf{u}}(\mathbf{k},t) = 0. \tag{4.2.10}$$

The evolution equations yield trajectories in the phase space spanned by the amplitudes $\hat{\mathbf{u}}(\mathbf{k},t)$, indexed by the countably infinite set of wavenumbers

$$\mathbf{k} = \hat{\mathbf{i}}\frac{2\pi n_1}{L} + \hat{\mathbf{j}}\frac{2\pi n_2}{L} + \hat{\mathbf{k}}\frac{2\pi n_3}{L} \tag{4.2.11}$$

for $n_i = 0, \pm 1, \pm 2, \ldots$. A PDE is generally an a priori infinite dimensional dynamical system. (Observe that this representation is a countably infinite set of ODEs because of the finite volume in which the PDE is defined.) A most surprising aspect of dissipative evolutions for continuum mechanical systems is that in some cases the dissipation can reduce the number of asymptotic degrees of freedom – the attractor dimension – to a finite number.

There are a number of definitions of the dimension of a set which generalize the usual integer dimensions to *fractal dimensions.* These definitions have several features in common, in particular (1) they agree for finite integer dimensional sets, and (2) the dimension of a set is at least as large as the dimension of any of its subsets. Two definitions with which we will work are the *Hausdorff* dimension and the *capacity* dimension. The Hausdorff dimension of a compact set A is defined

$$d_H = \inf\left\{d > 0 \middle| 0 = \lim_{r \searrow 0}\left(\inf\left\{\sum_{n=1}^{N} r_n^d \middle| A \subset \cup_{n=1}^{N} B_n,\right.\right.\right.$$

$$\left.\left.\left. B_n \text{ an open ball of radius } r_n \le r\right\}\right)\right\}, \qquad (4.2.12)$$

and the capacity dimension of A is defined

$$d_C = \limsup_{r \searrow 0} \frac{\log N(r)}{\log \frac{1}{r}}, \qquad (4.2.13)$$

where $N(r)$ is the minimal number of balls radii r required to cover A.

Likewise there are a variety of definitions of the "attractor" for a dynamical system but they too have several features in common. In this text we will generally be interested in estimates or bounds on the dimension of the *universal* or *global* attractor, $\mathcal{A}$, defined as follows. Consider all initial conditions in a ball B_ρ of radius ρ in the system phase space (in an appropriate metric). Let $B_\rho(t)$ be the image of B_ρ under the system's evolution for a time t. Then consider the set

$$\mathcal{A}_\rho = \cap_{t>0} B_\rho(t). \qquad (4.2.14)$$

This set consists of all points in phase space which can be arrived at starting from *all* times in the past at *some* point in B_ρ. Then we include points originating from ever larger initial conditions and define the global attractor

$$\mathcal{A} = \cup_{\rho>0} \mathcal{A}_\rho. \qquad (4.2.15)$$

So the universal attractor is the set of points in phase space that can be arrived at from some initial condition at an arbitrarily long time in the past. The more familiar properties associated with an attractor follow from this definition for $\mathcal{A}$:

- $\mathcal{A}$ is invariant under the evolution.
- The distance of any solution from $\mathcal{A}$ vanishes as $t \to \infty$.

Thus $\mathcal{A}$ contains all the asymptotic motion for the dynamical system. It is common to talk of "multiple attractors" for a dynamical system, each of which may in its own right be considered the attractor for initial conditions within its own basin of attraction. The notion of the global attractor corresponds to the union of all possible such dynamically invariant attracting sets and more. In particular $\mathcal{A}$ contains all possible structures such as fixed points (even completely unstable points known as "repellors"), limit cycles, etc., as well as their unstable manifolds.

A connection between the system dynamics and the attractor dimension is provided by the notion of the *Lyapunov* exponents via the *Kaplan-Yorke* formula. Roughly speaking, the Lyapunov exponents control the exponential growth or contraction of volume elements in phase space and the Kaplan-Yorke formula expresses the balance between volume growth and contraction realized on the attractor. The (global) Lyapunov exponents μ_i are determined according to the rule that the sum of the first n exponents gives the (largest possible) asymptotic exponential growth rate of n-volumes. If the sum of the first n exponents is negative, then all n-volumes decay to zero exponentially. Before giving the precise definition of the global Lyapunov exponents though, we give an indication of their relation to the attractor dimension through the Kaplan-Yorke formula.

The connection between the Lyapunov exponents and the capacity dimension of the attractor is heuristically established by the following argument. First suppose that the sum of the first $\tilde{n}$ exponents is negative, but the sum of the first $\tilde{n}-1$ is not. Then the attractor dimension should be less than $\tilde{n}$ because it cannot contain any $\tilde{n}$-dimensional subsets, and we presume that we may cover the attractor with $\tilde{n}$-dimensional balls of radius r. Suppose that it requires $N(r)$ such balls to cover the attractor. Then we let the system dynamics operate for a time t and observe that the $\tilde{n}$-dimensional balls of volume $\sim r^{\tilde{n}}$ evolve into $\tilde{n}$-dimensional ellipsoids of volume $\sim r^{\tilde{n}} \exp\{[\mu_1 + \ldots + \mu_{\tilde{n}}]t\}$ which still cover the attractor (see Figure 4.1). The smallest axis of the ellipsoids is $r\exp(\mu_{\tilde{n}}t) < r$, because $\mu_{\tilde{n}}$ is necessarily the most negative exponent. Then we cover each ellipsoid with balls of radius $r\exp(\mu_{\tilde{n}}t)$, which requires $\sim \exp\{[\mu_1 + \ldots + \mu_{\tilde{n}-1} + (1-\tilde{n})\mu_{\tilde{n}}]t\}$ smaller balls. So the total number of smaller balls required to cover the attractor is

$$N\left(r\exp(\mu_{\tilde{n}}t)\right) \sim \exp\{[\mu_1 + \ldots + \mu_{\tilde{n}-1} - (\tilde{n}-1)\mu_{\tilde{n}}]\, t\}\, N(r). \qquad (4.2.16)$$

According to formula (4.2.13) for the capacity dimension, then

$$\begin{aligned} d_C &\approx \limsup_{t\to\infty} \frac{\log\left[N\left(re^{\mu_{\tilde{n}}t}\right)\right]}{\log\left[\frac{1}{r}e^{-\mu_{\tilde{n}}t}\right]} \\ &= \tilde{n} - 1 + \frac{\mu_1 + \ldots + \mu_{\tilde{n}-1}}{-\mu_{\tilde{n}}}. \end{aligned} \qquad (4.2.17)$$

This is the Kaplan-Yorke formula. Note that according to the definition of $\tilde{n}$, the ratio of exponents in (4.2.17) satisfies

$$0 \leq \frac{\mu_1 + \ldots + \mu_{\tilde{n}-1}}{-\mu_{\tilde{n}}} < 1, \qquad (4.2.18)$$

so the Kaplan-Yorke formula generally yields a noninteger dimension

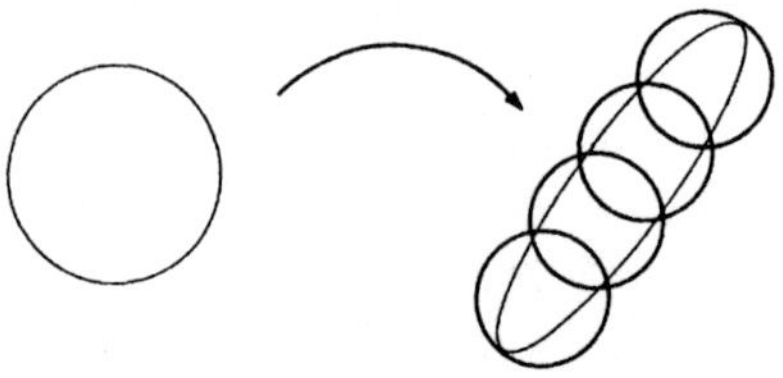

Fig. 4.1. An originally spherical volume evolves, under the system dynamics, into an ellipsoid which can be covered by balls of smaller amplitude.

between $\tilde{n}-1$ and $\tilde{n}$. We use it to define the *Lyapunov* dimension of the attractor:

$$d_L = \tilde{n} - 1 + \frac{\mu_1 + \ldots + \mu_{\tilde{n}-1}}{-\mu_{\tilde{n}}}, \tag{4.2.19}$$

where

$$\sum_{n=1}^{\tilde{n}-1} \mu_n \geq 0 \qquad \text{but} \qquad \sum_{n=1}^{\tilde{n}} \mu_n < 0. \tag{4.2.20}$$

The Lyapunov dimension differs conceptually from either the capacity or the Hausdorff dimensions in that its definition relies explicitly on the dynamics generating the attractor.

To give the precise definition of the global Lyapunov exponents, we will work in the context of an abstract dynamical system given by

$$\frac{d\mathbf{x}}{dt} = \mathbf{F}(\mathbf{x}), \tag{4.2.21}$$

where the dynamical variables $\mathbf{x}$ may be complex and of arbitrarily high dimension. Let $\mathbf{x}(t) = \{x_1(t), x_2(t), \ldots\}$ be a solution starting from initial condition $\mathbf{x}(0)$ and let $\mathbf{x}(t) + \delta\mathbf{x}(t) = \{x_1(t) + \delta x_1(t), x_2(t) + \delta x_2(t), \ldots\}$ be the solution starting an infinitesimal distance away at $\mathbf{x}(0) + \delta\mathbf{x}(0)$ as illustrated in Figure 4.2. The infinitesimal deviation $\delta\mathbf{x}(t)$ evolves according to the linearized evolution operator, linearized about the fully nonlinear solution $\mathbf{x}(t)$:

$$\frac{d\delta\mathbf{x}}{dt} = \mathbf{L}(\mathbf{x}(t)) \cdot \delta\mathbf{x}, \tag{4.2.22}$$

where the components of the linearized evolution operator are

$$L_{ij} = \frac{\partial F_i}{\partial x_j}. \tag{4.2.23}$$

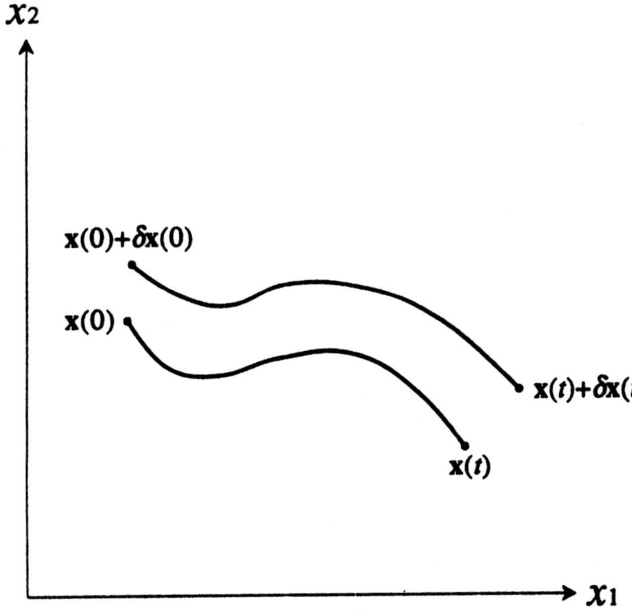

Fig. 4.2. Evolutions of neighboring trajectories.

The square of the length of the deviation, $|\delta\mathbf{x}|^2 = \delta\mathbf{x}^* \cdot \delta\mathbf{x}$, evolves as

$$\begin{aligned}\frac{d\,|\delta\mathbf{x}|^2}{dt} &= \frac{d\delta\mathbf{x}^*}{dt}\cdot\delta\mathbf{x} + \delta\mathbf{x}^*\cdot\frac{d\delta\mathbf{x}}{dt}\\ &= \delta\mathbf{x}^*\cdot\mathbf{L}^\dagger(\mathbf{x}(t))\cdot\delta\mathbf{x} + \delta\mathbf{x}^*\cdot\mathbf{L}(\mathbf{x}(t))\cdot\delta\mathbf{x}\\ &= \left[\delta\hat{\mathbf{x}}^*\cdot\left(\mathbf{L}^\dagger(\mathbf{x}(t)) + \mathbf{L}(\mathbf{x}(t))\right)\cdot\delta\hat{\mathbf{x}}\right]|\delta\mathbf{x}|^2, \qquad (4.2.24)\end{aligned}$$

where $\delta\hat{\mathbf{x}}(t) = \delta\mathbf{x}/|\delta\mathbf{x}|$ is the unit vector in the direction of $\delta\mathbf{x}(t)$, and $\mathbf{L}^\dagger$ is the Hermitian transpose (conjugate transpose) of the matrix $\mathbf{L}$. Denote the Hermitian part of the linearized evolution operator $\mathbf{L}^H = \frac{1}{2}(\mathbf{L}+\mathbf{L}^\dagger)$. It is a time dependent linear operator depending on the initial condition $\mathbf{x}(0)$ – indeed, it depends on the ensuing trajectory $\mathbf{x}(t)$ – and we denote this dependence by writing $\mathbf{L}^H = \mathbf{L}^H(t;\mathbf{x}(0))$.

Define the matrix $\mathbf{P}^{(1)}(t) = \delta\hat{\mathbf{x}}(t)\delta\hat{\mathbf{x}}(t)^*$, with components $P_{ij} = (\delta\hat{x})_i(\delta\hat{x})_j^*$, which projects vectors along the direction of $\delta\mathbf{x}(t)$. Then the length of the deviation satisfies

$$\begin{aligned}\frac{d\,|\delta\mathbf{x}|}{dt} &= \left[\delta\hat{\mathbf{x}}^{*}\cdot\mathbf{L}^{H}(t;\mathbf{x}(0))\cdot\delta\hat{\mathbf{x}}\right]|\delta\mathbf{x}| \\ &= \left(\delta\hat{x}_i^{*}L_{ij}^{H}\delta\hat{x}_j\right)|\delta\mathbf{x}| \\ &= L_{ij}^{H}P_{ji}|\delta\mathbf{x}| \\ &= Tr\left[\mathbf{L}^{H}(t;\mathbf{x}(0))\,\mathbf{P}^{(1)}(t)\right]|\delta\mathbf{x}|, \end{aligned} \tag{4.2.25}$$

where Tr denotes the trace operation. This differential equation may be integrated to yield

$$\begin{aligned}|\delta\mathbf{x}(t)| &= |\delta\mathbf{x}(0)|\exp\left\{\int_0^t Tr\left[\mathbf{L}^{H}(t';\mathbf{x}(0))\,\mathbf{P}^{(1)}(t')\right]dt'\right\} \\ &= |\delta\mathbf{x}(0)|\exp\left\{\left[\frac{1}{t}\int_0^t Tr\left[\mathbf{L}^{H}(t';\mathbf{x}(0))\,\mathbf{P}^{(1)}(t')\right]dt'\right]\times t\right\}. \end{aligned} \tag{4.2.26}$$

Thus the deviation grows or shrinks exponentially, asymptotically as $t \to \infty$, depending on the sign of the time averaged trace of the Hermitian part (or symmetric part, if everything is real) of the linearized evolution operator composed with the projection onto the span of the deviation. Note also that $Tr\left[\mathbf{L}^{H}(t;\mathbf{x}(0))\,\mathbf{P}^{(1)}(t)\right] = \Re\left[Tr(\mathbf{L}(t;\mathbf{x}(0))\,\mathbf{P}^{(1)}(t))\right]$, i.e., it's just the real part of the trace of $\mathbf{L}$ composed with the projection operator $\mathbf{P}^{(1)}$.

If the limit exists, the first Lyapunov exponent is defined as

$$\lim_{t\to\infty}\frac{1}{t}\int_0^t Tr\left[\mathbf{L}^{H}(t';\mathbf{x}(0))\,\mathbf{P}^{(1)}(t')\right]dt', \tag{4.2.27}$$

which generally depends on the starting point $\mathbf{x}(0)$. Sometimes the exponent is the same for almost all initial points, in which case a unique exponent characterizes the dynamics. A positive Lyapunov exponent indicates that small initial deviations grow exponentially, asymptotically in time as $t \to \infty$. This is usually the definition of *chaos* in a dynamical system, as it indicates a sensitive dependence of the evolution on the precise initial conditions, and hence the essential impossibility of long term prediction.

The first global Lyapunov exponent μ_1 is the largest possible asymptotic exponential growth rate of infinitesimal one-dimensional displacements on the global attractor. Referring to the expression above, we define

$$\mu_1 = \limsup_{t\to\infty}\ \sup_{\mathbf{x}(0)\in\mathscr{A}}\ \sup_{\mathbf{P}^{(1)}(0)}\left\{\frac{1}{t}\int_0^t Tr\left[\mathbf{L}^{H}(t';\mathbf{x}(0))\,\mathbf{P}^{(1)}(t')\right]dt'\right\}. \tag{4.2.28}$$

A positive global Lyapunov exponent is consistent with chaos, but more is really necessary to ensure dynamical chaos.

The sum of the first n global Lyapunov exponents is defined, in a similar manner, as the largest possible asymptotic exponential growth rate of infinitesimal n-volumes. To make this explicit, consider the infinitesimal n-volume spanned by the infinitesimal vectors $\delta\mathbf{x}^{(1)}(0), \ldots, \delta\mathbf{x}^{(n)}(0)$. Each edge of the volume is evolved by the linearized evolution. That is, $\delta\mathbf{x}^{(i)}(t)$ satisfies

$$\frac{d\delta\mathbf{x}^{(i)}}{dt} = \mathbf{L}[\mathbf{x}(t)] \cdot \delta\mathbf{x}^{(i)} \tag{4.2.29}$$

with initial condition $\delta\mathbf{x}^{(i)}(0)$. The volume element itself changes both in orientation and size as the system develops, as illustrated in Figure 4.3. We denote the time-dependent volume as the "n-form" $\delta\mathbf{x}^{(1)}(t) \wedge \ldots \wedge \delta\mathbf{x}^{(n)}(t)$, whose magnitude is the volume $V_n(t) = \left|\delta\mathbf{x}^{(1)}(t) \wedge \ldots \wedge \delta\mathbf{x}^{(n)}(t)\right|$, where the n-form norm is

$$\left|\mathbf{v}^{(1)} \wedge \ldots \wedge \mathbf{v}^{(n)}\right|^2 = \det \begin{pmatrix} \mathbf{v}^{(1)*} \cdot \mathbf{v}^{(1)} & \mathbf{v}^{(1)*} \cdot \mathbf{v}^{(2)} & \ldots & \mathbf{v}^{(1)*} \cdot \mathbf{v}^{(n)} \\ \mathbf{v}^{(2)*} \cdot \mathbf{v}^{(1)} & \mathbf{v}^{(2)*} \cdot \mathbf{v}^{(2)} & \ldots & \mathbf{v}^{(2)*} \cdot \mathbf{v}^{(n)} \\ \vdots & \vdots & \vdots & \vdots \\ \mathbf{v}^{(n)*} \cdot \mathbf{v}^{(1)} & \mathbf{v}^{(n)*} \cdot \mathbf{v}^{(2)} & \ldots & \mathbf{v}^{(n)*} \cdot \mathbf{v}^{(n)} \end{pmatrix}. \tag{4.2.30}$$

That is

$$V_n(t)^2 = \det \mathbf{M}(t), \tag{4.2.31}$$

where

$$M_{ij}(t) = \delta\mathbf{x}^{(i)}(t)^* \cdot \delta\mathbf{x}^{(j)}(t) \tag{4.2.32}$$

is an $n \times n$ Hermitian matrix. The eigenvalues of $\mathbf{M}$ are real and its determinant is the product of those real eigenvalues. Recalling the identity $\log\det[\mathbf{M}] = Tr[\log\mathbf{M}]$, we have

$$2\frac{d\log V_n}{dt} = \frac{d}{dt}\log\det\mathbf{M} = \frac{d}{dt}Tr(\log\mathbf{M}) = Tr\left(\mathbf{M}^{-1}\frac{d\mathbf{M}}{dt}\right). \tag{4.2.33}$$

To evaluate this trace, we introduce a time-dependent orthonormal set of basis vectors $\boldsymbol{\phi}^{(1)}(t), \ldots, \boldsymbol{\phi}^{(n)}(t)$ which span the span of $\delta\mathbf{x}^{(1)}(t) \ldots$, $\delta\mathbf{x}^{(n)}(t)$. (The $\boldsymbol{\phi}$'s can be constructed from the $\delta\mathbf{x}$'s by a Gramm-Schmidt procedure.) Define the matrix $\mathbf{m}(t)$, of the components of the $\delta\mathbf{x}$'s, by

$$m_{ij} = \boldsymbol{\phi}^{(i)*} \cdot \delta\mathbf{x}^{(j)}, \tag{4.2.34}$$

so that

$$\mathbf{M} = \mathbf{m}^\dagger\mathbf{m} \qquad \mathbf{M}^{-1} = \mathbf{m}^{-1}(\mathbf{m}^\dagger)^{-1}. \tag{4.2.35}$$

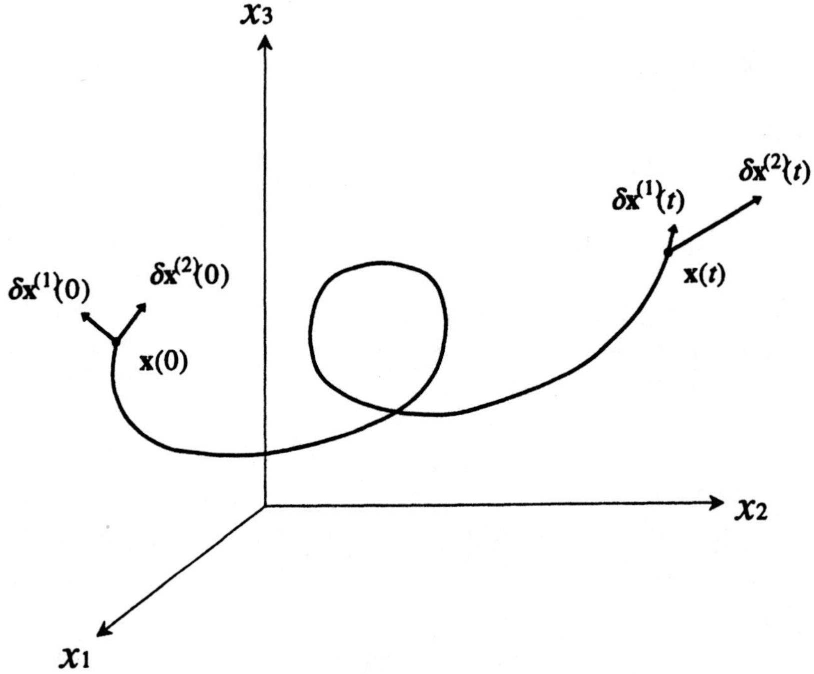

Fig. 4.3. Originally orthogonal infinitesimal vectors, $\delta\mathbf{x}^{(1)}(0)$ and $\delta\mathbf{x}^{(2)}(0)$, evolve into $\delta\mathbf{x}^{(1)}(t)$ and $\delta\mathbf{x}^{(2)}(t)$ which are no longer necessarily orthogonal.

The matrix elements of the linearized evolution operator with respect to this basis are

$$a_{kl} = \boldsymbol{\phi}^{(k)*} \cdot \mathbf{L} \cdot \boldsymbol{\phi}^{(l)}. \tag{4.2.36}$$

Then

$$\begin{aligned}
\frac{dM_{ij}}{dt} &= \frac{d\delta\mathbf{x}^{(i)*}}{dt} \cdot \delta\mathbf{x}^{(j)} + \delta\mathbf{x}^{(i)*} \cdot \frac{d\delta\mathbf{x}^{(j)}}{dt} \\
&= \delta\mathbf{x}^{(i)*} \cdot \mathbf{L}^{\dagger} \cdot \delta\mathbf{x}^{(j)} + \delta\mathbf{x}^{(i)*} \cdot \mathbf{L} \cdot \delta\mathbf{x}^{(j)} \\
&= \sum_{k,l=1}^{n} \left\{ \delta\mathbf{x}^{(i)*} \cdot \boldsymbol{\phi}^{(k)} \boldsymbol{\phi}^{(k)*} \cdot (\mathbf{L} + \mathbf{L}^{\dagger}) \cdot \boldsymbol{\phi}^{(l)} \boldsymbol{\phi}^{(l)*} \cdot \delta\mathbf{x}^{(j)} \right\} \\
&= \sum_{k,l=1}^{n} \left(m_{ik}^{*} (a_{kl}^{*} + a_{kl}) m_{lj} \right).
\end{aligned} \tag{4.2.37}$$

In the more compact notation, denoting by $\mathbf{a}$ the matrix with elements

a_{ij}, this is

$$\frac{d\mathbf{M}}{dt} = \mathbf{m}^{\dagger}\mathbf{a}^{\dagger}\mathbf{m} + \mathbf{m}^{\dagger}\mathbf{a}\mathbf{m} = \mathbf{m}^{\dagger}(\mathbf{a}^{\dagger} + \mathbf{a})\mathbf{m}. \quad (4.2.38)$$

Putting together (4.2.33), (4.2.35), and (4.2.38), and using the cyclic properties of the trace, the logarithm of the volume element is seen to satisfy

$$\begin{aligned} 2\frac{d \log V_n}{dt} &= Tr\left(\mathbf{m}^{-1}(\mathbf{m}^{\dagger})^{-1}\mathbf{m}^{\dagger}(\mathbf{a}^{\dagger} + \mathbf{a})\mathbf{m}\right) \\ &= Tr(\mathbf{a}^{\dagger} + \mathbf{a}) \\ &= 2Tr(\mathbf{a}^{H}), \end{aligned} \quad (4.2.39)$$

where $\mathbf{a}^H$ is the Hermitian part of $\mathbf{a}$.

Define the projection operator, $\mathbf{P}^{(n)}(t)$, onto the span of $\delta\mathbf{x}^{(1)}(t), \ldots, \delta\mathbf{x}^{(n)}(t)$:

$$\mathbf{P}^{(n)}(t) = \sum_{i=1}^{n} \boldsymbol{\phi}^{(i)}\boldsymbol{\phi}^{(i)*}. \quad (4.2.40)$$

Then the trace in (4.2.39) is

$$\begin{aligned} Tr\left(\mathbf{a}^{H}\right) &= \sum_{i=1}^{n} \boldsymbol{\phi}^{(i)*} \cdot \frac{1}{2}(\mathbf{L} + \mathbf{L}^{\dagger}) \cdot \boldsymbol{\phi}^{(i)} \\ &= Tr\left[\mathbf{L}^{H}\mathbf{P}^{(n)}(t)\right] \\ &= \Re\left(Tr\left[\mathbf{L}\mathbf{P}^{(n)}(t)\right]\right). \end{aligned} \quad (4.2.41)$$

Solving for the time-dependent volume we find

$$\begin{aligned} V_n(t) &= V_n(0)\exp\left\{\int_0^t Tr\left[\mathbf{L}^{H}\left(t';\mathbf{x}(0)\right)\mathbf{P}^{(n)}(t')\right] dt'\right\} \\ &= V_n(0)\exp\left\{\left[\frac{1}{t}\int_0^t \Re\left\{Tr\left[\mathbf{L}\left(t';\mathbf{x}(0)\right)\mathbf{P}^{(n)}(t')\right]\right\} dt'\right] \times t\right\}. \end{aligned} \quad (4.2.42)$$

The sum of the first n Lyapunov exponents is defined as the largest possible asymptotic exponential growth rate of volume elements around the attractor:

$$\mu_1 + \ldots + \mu_n = \limsup_{t\to\infty} \sup_{\mathbf{x}(0)\in\mathscr{A}} \sup_{\mathbf{P}^{(n)}(0)} \left\{\frac{1}{t}\int_0^t Tr\left[\mathbf{L}^{H}\left(t';\mathbf{x}(0)\right)\mathbf{P}^{(n)}(t')\right] dt'\right\}. \quad (4.2.43)$$

The association of the Lyapunov and capacity dimensions given in the discussion around (4.2.16) and (4.2.17) above is not a rigorous relation. What is known is due to a theorem of Constantin and Foias:[1] the Lyapunov dimension, with the global Lyapunov exponents as defined in (4.2.43), is an upper bound on the Hausdorff dimension of the universal attractor. This theorem provides the rigorous basis for our earlier assertion that if all n-dimensional volume elements in the phase space vanish as $t \to \infty$, then the attractor dimension is less than n. A most remarkable feature of Constantin and Foias' theorem is that it holds also for a priori infinite dimensional dynamical systems such as PDEs.

We will approach the problem of producing estimates – upper and lower bounds – for the universal attractor dimension for the Navier-Stokes equations from an operational point of view. As far as upper bounds are concerned we will appeal to the theorem of Constantin and Foias and conclude that if an attractor contains no n-dimensional subsets, i.e., if the sum of the first n global Lyapunov exponents is negative, then its dimension is less than or equal to n. Lower bounds on the attractor dimension will be provided by the dimension of unstable manifolds contained in the attractor. For example, the dimension of the unstable manifold of a fixed point is the number of positive eigenvalues of the linearized evolution operator, linearized about the fixed point. The ultimate goal of this kind of analysis will be to derive, directly from the Navier-Stokes equations, estimates of the number of degrees of freedom identified as the dimension of the universal attractor, as a function of nondimensional quantities like Reynolds, Grashof, or Rayleigh numbers. This task is taken up in Chapter 9. In order to illustrate these ideas, as well as the techniques used in their implementation, in the next section we develop the particular example of the Lorenz equations.

4.3 The Lorenz system

In this section we use the Lorenz equations to illustrate some of the methods that have been described and developed so far. After showing how the Lorenz equations are derived as a truncated model of thermal convection from the Boussinesq equations, we apply both linear and nonlinear stability arguments to determine some features of the system dynamics. Related analytical techniques, e.g., analyses via Lyapunov functions, are then used to locate the attractor when it is more complicated than just fixed points. The Lyapunov exponents are then estimated and used to derive a bound on the Lyapunov dimension of the global attractor.

[1] See reference section 4.4.

The Lorenz equations define the evolution of 3 dynamical variables, $X(t), Y(t)$, and $Z(t)$:

$$\frac{dX}{dt} = -\sigma X + \sigma Y \tag{4.3.1}$$

$$\frac{dY}{dt} = rX - Y - XZ \tag{4.3.2}$$

$$\frac{dZ}{dt} = XY - bZ, \tag{4.3.3}$$

where σ, b and r are positive parameters. One of the classic models of nonlinear dynamics and chaos, these equations were originally derived in a modal truncation of the Boussinesq equations for thermal convection.[1] There, σ corresponds to the Prandtl number, b is a geometric parameter, and $r = Ra/Ra_c$ is the Rayleigh number in units of the critical Rayleigh number.

To see how these equations arise, consider the nondimensional form of the Boussinesq equations for $2d$ convection (recall (2.1.31)-(2.1.33)) in the $x - z$ plane,

$$\frac{\partial \mathbf{u}}{\partial t} + \mathbf{u} \cdot \nabla \mathbf{u} + \nabla p = \sigma \Delta \mathbf{u} + \hat{\mathbf{k}} \sigma Ra\, \theta, \tag{4.3.4}$$

$$\nabla \cdot \mathbf{u} = 0, \tag{4.3.5}$$

$$\frac{\partial \theta}{\partial t} + \mathbf{u} \cdot \nabla \theta = \Delta \theta + u_3, \tag{4.3.6}$$

where θ is the deviation of the temperature from the linear conduction profile. The stream function, ψ, for the $2d$ velocity field is defined by

$$u_1 = -\frac{\partial \psi}{\partial z}, \qquad u_3 = \frac{\partial \psi}{\partial x}. \tag{4.3.7}$$

The (stress-free) boundary conditions on the bottom plate at $z = 0$ and the top plate at $z = 1$ are

$$\psi = 0, \qquad \Delta \psi = 0, \qquad \theta = 0. \tag{4.3.8}$$

Periodic boundary conditions are imposed in the x-direction, on a length

[1] The Lorenz equations, although in themselves fascinating equations, and although they do mirror certain features of the dynamics, are not a very good representation of the $2d$ Boussinesq equations for arbitrarily large values of Ra. They are nevertheless interesting to pure and applied mathematicians because of the chaotic dynamics they display and because of the important role they played in early developments in the theory of chaos.

$2\pi/k$. Introducing the "Galerkin" truncation

$$\psi = \frac{\sqrt{2}}{\pi}\left(\frac{k^2+\pi^2}{k}\right) X \sin(kx)\sin(\pi z) \tag{4.3.9}$$

$$\theta = \frac{\sqrt{2}}{\pi r} Y \cos(kx)\sin(\pi z) - \frac{1}{\pi r} Z \sin(2\pi z) \tag{4.3.10}$$

and rescaling time by $t \to 3\pi^2 t$, we find (after substitution into the Boussinesq equations and neglect of all modes beyond those in (4.3.9) and (4.3.10)) that the mode amplitudes X, Y, and Z satisfy the Lorenz equations (4.3.1) - (4.3.3) with the geometric parameter

$$b = \frac{4\pi^2}{k^2+\pi^2}, \tag{4.3.11}$$

and where the critical Rayleigh number for wavenumber k is

$$Ra_c = \frac{(k^2+\pi^2)^3}{k^2}. \tag{4.3.12}$$

The usual parameter values are $\sigma = 10$ (a Prandtl number appropriate for the atmosphere) and $k = \pi\sqrt{2}$ (corresponding to the smallest possible value of Ra_c), which gives $b = 8/3$. The renormalized Rayleigh number r is then the natural control parameter, with σ and b secondary parameters.

The Lorenz equations describe a dissipative dynamical system for all values of r because the divergence of the flow field is always negative:

$$\begin{aligned}\frac{\partial}{\partial X}(-\sigma X + \sigma Y) \quad &+ \quad \frac{\partial}{\partial Y}(rX - Y - XZ) + \frac{\partial}{\partial Z}(XY - bZ) \\ &= \quad -\sigma - 1 - b < 0.\end{aligned} \tag{4.3.13}$$

Hence 3-dimensional volumes in the phase space contract to zero at a uniform exponential rate, and the system's attractor is necessarily of dimension less than three.

The first step in the analysis is to locate and determine the stability of the system's fixed points. Fixed points of the Lorenz equations satisfy the nonlinear simultaneous equations obtained from (4.3.1) - (4.3.3) by setting the time derivatives to zero. One solution, valid for all parameter values, is the origin:

$$(X, Y, Z) = (0, 0, 0). \tag{4.3.14}$$

This solution corresponds to the pure conduction state (no flow) in the convection problem. Two other "nontrivial" solutions exist for $r \geq 1$:

$$(X, Y, Z) = \left(\pm\sqrt{b(r-1)}, \pm\sqrt{b(r-1)}, r-1\right). \tag{4.3.15}$$

The condition $r \geq 1$ corresponds[1] to $Ra > Ra_c$.

The linear stability of the origin is investigated by considering the evolution of infinitesimal disturbances,

$$\frac{d}{dt}\begin{pmatrix} \delta X \\ \delta Y \\ \delta Z \end{pmatrix} = \begin{pmatrix} -\sigma & \sigma & 0 \\ r & -1 & 0 \\ 0 & 0 & -b \end{pmatrix}\begin{pmatrix} \delta X \\ \delta Y \\ \delta Z \end{pmatrix}. \tag{4.3.16}$$

The time dependence of solutions to these linear equations are superpositions of functions of the form $e^{\lambda t}$, where λ is an eigenvalue of the evolution operator. If the real parts of the eigenvalues are all negative, then stability against small perturbations is indicated. If at least one eigenvalue has a positive real part, then the origin is unstable to an infinitesimal perturbation. The eigenvalues satisfy the characteristic equation

$$0 = (\lambda + b)\left(\lambda^2 + (\sigma+1)\lambda - \sigma(r-1)\right). \tag{4.3.17}$$

Hence $\lambda = -b$ is always an eigenvalue, and the other two are

$$\lambda_\pm = -\frac{\sigma+1}{2} \pm \sqrt{\left(\frac{\sigma+1}{2}\right)^2 + \sigma(r-1)}. \tag{4.3.18}$$

These latter two eigenvalues are always real, and we see that λ_+ crosses from negative to positive as r passes through 1. Hence the origin is linearly stable for $r < 1$, and it is unstable for $r > 1$. This is precisely the loss of stability of the origin $(0,0,0)$ for $Ra > Ra_c$.

More can be shown: When $r < 1$ the origin is actually nonlinearly stable. To see this we appeal to a nonlinear stability analysis based on the "energy" function

$$H(X,Y,Z) = \frac{1}{2}(X^2+\sigma Y^2+\sigma Z^2) = \frac{1}{2}\begin{pmatrix} X & \sqrt{\sigma}Y & \sqrt{\sigma}Z \end{pmatrix}\begin{pmatrix} X \\ \sqrt{\sigma}Y \\ \sqrt{\sigma}Z \end{pmatrix}. \tag{4.3.19}$$

The origin is the unique global minimum of H. Using the equations of

[1] These two solutions represent steady convection rolls with two senses of rotation.

motion, we see that H evolves according to

$$\frac{dH}{dt} = \begin{pmatrix} X & \sqrt{\sigma}Y & \sqrt{\sigma}Z \end{pmatrix} \begin{pmatrix} \sigma & -\frac{\sqrt{\sigma}}{2}(r-1) & 0 \\ -\frac{\sqrt{\sigma}}{2}(r-1) & 1 & 0 \\ 0 & 0 & b \end{pmatrix} \begin{pmatrix} X \\ \sqrt{\sigma}Y \\ \sqrt{\sigma}Z \end{pmatrix}$$

$$\leq 2\lambda_{max} H, \tag{4.3.20}$$

where λ_{max} is the largest (real) eigenvalue of the (Hermitian) matrix in (4.3.20) above. According to Gronwall's lemma (see Lemma 2.1),

$$H(t) \leq H(0) \exp(2\lambda_{max} t) \tag{4.3.21}$$

so if λ_{max} is negative then H decays to zero exponentially and the origin is nonlinearly stable.

The eigenvalues of the matrix in (4.3.20) satisfy the characteristic equation

$$0 = (\lambda + b)\left(\lambda^2 + (\sigma+1)\lambda + \frac{\sigma}{4}(3+r)(1-r)\right), \tag{4.3.22}$$

with roots

$$\lambda = -b < 0 \tag{4.3.23}$$

and

$$\lambda_{\pm} = -\frac{\sigma+1}{2} \pm \sqrt{\left(\frac{\sigma+1}{2}\right)^2 + \frac{\sigma}{4}(3+r)(r-1)}. \tag{4.3.24}$$

Both λ_+ and λ_- are negative for $r < 1$, with λ_+ passing through 0 to positive values as r increases through 1.

Combining the results of these linear and nonlinear stability analyses, we may draw a number of conclusions. First, the origin is the global attractor for $r < 1$. Second, it is unstable for $r > 1$, and third, the origin has a one-dimensional unstable manifold for values of $r > 1$. The origin undergoes a supercritical pitchfork bifurcation at the critical control parameter value $r = 1$, where the new nontrivial fixed points, given in (4.3.15), come into play.

The linear stability of the nontrivial fixed points is determined by the linearized evolution equations about these fixed points

$$\frac{d}{dt}\begin{pmatrix} \delta X \\ \delta Y \\ \delta Z \end{pmatrix} = \begin{pmatrix} -\sigma & \sigma & 0 \\ 1 & -1 & -c \\ c & c & -b \end{pmatrix} \begin{pmatrix} \delta X \\ \delta Y \\ \delta Z \end{pmatrix}, \tag{4.3.25}$$

where $c = \pm\sqrt{b(r-1)}$. Solutions are superpositions of terms of the form

$e^{\lambda t}$ where the eigenvalues λ are roots of

$$0 = \lambda^3 + (\sigma + 1 + b)\lambda^2 + b(\sigma + r)\lambda + 2\sigma b(r - 1). \tag{4.3.26}$$

The roots of the cubic equation are not very simple, but we may extract some essential information without resorting to writing them down explicitly. When the nontrivial fixed points come into existence at $r = 1$, (4.3.26) factors simply and the eigenvalues are

$$\lambda = 0, \qquad \text{or} \qquad \lambda = -b, \qquad \text{or} \qquad \lambda = -(\sigma + 1), \tag{4.3.27}$$

indicating (marginal) stability. For values of r slightly above unity, the two negative roots remain negative and the zero eigenvalue becomes

$$\lambda \approx -\frac{2\sigma(r-1)}{\sigma + 1} < 0, \tag{4.3.28}$$

indicating stability. We conclude that the nontrivial fixed points are stable for values of r not too far above 1, although they may lose stability as r increases.

Because all the coefficients of this characteristic polynomial (4.3.26) are positive (in the relevant parameter regime where $\sigma > 0, b > 0$, and $r > 1$), there are no positive real roots to the equation. Hence if the nontrivial fixed points are to become unstable as r is increased, it must be as a pair of complex conjugate roots passes the imaginary axis. If this does occur at some critical value $r = r_c$, then at that point two of the roots are purely imaginary and of opposite sign. The trace of the matrix is the sum of the roots, so at $r = r_c$ the third eigenvalue is $-(\sigma + 1 + b)$. This consideration determines r_c. Demanding that $-(\sigma+1+b)$ be a root fixes r_c in terms of σ and b:

$$r_c = \frac{\sigma(\sigma + b + 3)}{\sigma - b - 1}. \tag{4.3.29}$$

This expression for r_c is positive and finite only when $\sigma > 1+b$, in which case it is always greater than one, leading us to deduce that

- for $\sigma \leq 1 + b$, the nontrivial fixed points remain stable for all $r > 1$,
- for $\sigma > 1 + b$, instability occurs as r is increased through r_c.

If they are unstable, then the nontrivial fixed points are unstable foci, with nearby solutions spiraling away on a two-dimensional unstable manifold.[1]

[1] This instability corresponds to a breakdown of steady convection rolls in an oscillatory manner.

Above the instability at r_c, (presuming that $\sigma > 1+b$ so $1 < r_c < \infty$) the Lorenz equations possess no stable fixed points. Finding and describing the system's attractor then requires other considerations. For example, we know that the global attractor contains no three-dimensional subsets. It contains the fixed point at the origin and its one-dimensional unstable manifold, as well as the two nontrivial fixed points along with their two-dimensional unstable manifolds. The global attractor is therefore of dimension between two and three.

Using the method of Lyapunov functionals we may locate the attractor approximately in phase space. For example, the attractor of the Lorenz equations is always contained within a sphere centered on the point $(0, 0, r+\sigma)$. To see this, define the function

$$K(X,Y,Z) = X^2 + Y^2 + (Z - r - \sigma)^2, \tag{4.3.30}$$

and consider its evolution as determined by the Lorenz equations:

$$\frac{1}{2}\frac{dK}{dt} = -\sigma X^2 - Y^2 - bZ^2 + b(r+\sigma)Z. \tag{4.3.31}$$

If X, Y, and Z are all very large, then K will be a decreasing function of time, and the trajectories will move closer to the point $(0, 0, r+\sigma)$. We can bound the radius of a sphere into which all solutions flow by rewriting the right-hand side of (4.3.31) as

$$\begin{aligned} \frac{1}{2}\frac{dK}{dt} &= -\sigma X^2 - Y^2 - \frac{1}{2}b(Z - r - \sigma)^2 - \frac{1}{2}bZ^2 + \frac{1}{2}b(r+\sigma)^2 \\ &\leq -\min(\sigma, 1, b/2)K + \frac{1}{2}b(r+\sigma)^2. \end{aligned} \tag{4.3.32}$$

It is apparent that if K is large enough, then its time derivative is negative. Gronwall's lemma implies that, as $t \to \infty$,

$$\overline{\lim}_{t\to\infty} K(t) \leq \frac{b(r+\sigma)^2}{\min(2\sigma, 2, b)}, \tag{4.3.33}$$

where the overbar refers to a limit supremum additionally over all initial conditions. A slightly more delicate analysis yields an improved bound on the radius. The largest value that K can realize asymptotically as $t \to \infty$ is no larger than its largest value inside the solid ellipse defined by the condition that the right-hand side of (4.3.31) is positive. This is because if K is larger than this maximum, then dK/dt is surely negative, and the system will evolve to decrease K. If K is less than or equal to this maximum, then it cannot increase through the surface of the sphere because $dK/dt \le 0$ there. The radius of this sphere is the solution K_{max} of the constrained maximization problem,

$$K_{max} = \max\left\{K(X,Y,Z) \mid \sigma X^2 + Y^2 + bZ^2 - b(r+\sigma)Z \ge 0\right\}. \quad (4.3.34)$$

Using the method of Lagrange multipliers, we find

$$\overline{\lim}_{t\to\infty} K(t) \le K_{max} = \begin{cases} (r+\sigma)^2 & \text{for } b \le 2, \\ \frac{b^2(r+\sigma)^2}{4(b-1)} & \text{for } b \ge 2. \end{cases} \quad (4.3.35)$$

All solutions which start outside the sphere of radius $\sqrt{K_{max}}$ about the point $(0,0,r+\sigma)$ approach it, and those which start inside will remain inside in the subsequent evolution. For this reason this sphere is referred to as an "absorbing ball."

Locating the attractor inside a sphere is, however, a rather featureless characterization. By considering some other Lyapunov functionals though, we can bound it more definitely. Define

$$J(Y,Z) = Y^2 + (Z-r)^2, \quad (4.3.36)$$

which evolves according to

$$\frac{1}{2}\frac{dJ}{dt} = -Y^2 - bZ^2 + rbZ. \quad (4.3.37)$$

Using the method of Lagrange multipliers to find the maximum of J inside the solid ellipse defined by the positivity of the right-hand side of (4.3.37), we conclude that as $t \to \infty$,

$$\overline{\lim}_{t\to\infty} J(t) \le J_{max} = \begin{cases} r^2 & \text{for } b \le 2, \\ \frac{b^2 r^2}{4(b-1)} & \text{for } b \ge 2. \end{cases} \quad (4.3.38)$$

This places the attractor inside the intersection of the absorbing ball defined by $K \le K_{max}$ and the cylinder of radius $\sqrt{J_{max}}$, with axis parallel to the x-axis, and passing through the point $(0,0,r)$.

Now define

$$I(X,Z) = X^2 - 2\sigma Z, \quad (4.3.39)$$

which satisfies

$$\begin{aligned} \frac{1}{2}\frac{dI}{dt} &= -\sigma X^2 + \sigma bZ \\ &= -\sigma I - \sigma(2\sigma - b)Z. \end{aligned} \quad (4.3.40)$$

The range of Z on the attractor is bounded by the surface of the cylinder $J \le J_{max}$. That is, on the attractor

$$Z_{min} = r - \sqrt{J_{max}} \le Z \le r + \sqrt{J_{max}} = Z_{max}. \quad (4.3.41)$$

The time asymptotic motion for the parameter regime where $2\sigma \geq b$ (which is implied by the condition $\sigma > b+1$ necessary for the interesting case of $0 < r_c < \infty$) satisfies the pair of differential inequalities

$$-I - (2\sigma - b)Z_{max} \leq \frac{1}{2\sigma}\frac{dI}{dt} \leq -I - (2\sigma - b)Z_{min}. \qquad (4.3.42)$$

Gronwall's lemma (see Lemma 2.1) then bounds I for large t on the attractor according to

$$-(2\sigma - b)Z_{max} \leq I \leq -(2\sigma - b)Z_{min}, \qquad (4.3.43)$$

which in turn implies that Z is sandwiched between two parabolic sheets lying parallel to the y-axis:

$$\frac{1}{2\sigma}X^2 + \left(1 - \frac{b}{2\sigma}\right) Z_{min} \leq Z \leq \frac{1}{2\sigma}X^2 + \left(1 - \frac{b}{2\sigma}\right) Z_{max}. \qquad (4.3.44)$$

A slight but important improvement in the bounds is obtained by considering the ellipsoid

$$G(X,Y,Z) = r^2X^2 + \sigma Y^2 + \sigma\,[Z - r(r-1)]^2. \qquad (4.3.45)$$

Differentiating with respect to time gives

$$\begin{aligned}\frac{1}{2}\frac{dG}{dt} &= r^2X\,(-\sigma X + \sigma Y) + \sigma Y\,(-Y + (r-Z)X) \\ &\quad + \sigma\,[Z - r(r-1)]\,(-bZ + XY),\end{aligned} \qquad (4.3.46)$$

so rearranging and completing the square, we have

$$\frac{1}{2}\frac{dG}{dt} = -\sigma\,(rX - Y)^2 - \sigma b Z^2 + \sigma b r(r-1)Z. \qquad (4.3.47)$$

The constrained maximization problem is similar to that for the sphere and the cylinder and is

$$G_{max} = \max\{G(X,Y,Z) | \sigma(rX - Y)^2 + \sigma b Z^2 - \sigma b r(r-1)Z \geq 0)\}. \qquad (4.3.48)$$

After a little work we find that G decreases if it is greater than

$$G_{max} = \sigma r^2(r-1)^2. \qquad (4.3.49)$$

On top of the sphere, the cylinder and the two parabolic sheets, this result additionally places the attractor inside the ellipsoid defined by

$$G(X,Y,Z) \leq G_{max}. \qquad (4.3.50)$$

In particular, this shows that on the global attractor, $Z \geq 0$.

To illustrate these estimates in a particular case, Figure 4.4 shows three projections of a trajectory near the global attractor plotted together with the bounds on the variables implied by the analyses above. The parameter values used here are $\sigma = 10, b = 8/3$ and $r = 28$, corresponding to chaotic dynamics. The trajectory in Figure 4.4 starts on the unstable manifold of the origin, and moves first toward one of the nontrivial fixed points. For these values of σ and b, $r_c = 24.7\ldots$, so those fixed points are unstable. The trajectory then spirals outward until it comes near the stable manifold of the other nontrivial fixed point. This point too is unstable and the trajectory orbits outward[1] until it flips over to the other "wing." The global attractor for these parameter values is a fractal set of dimension between 2 and 3.

The limitations on the dynamics on the global attractor deduced above can be used in conjunction with the methods of the last section to develop a more precise estimate of the attractor's dimension. The linearized evolution operator about the trajectory $\mathbf{X}(t) = (X(t), Y(t), Z(t))$ starting at $\mathbf{X}(0)$ is

$$\frac{d}{dt}\begin{pmatrix} \delta X \\ \delta Y \\ \delta Z \end{pmatrix} = \begin{pmatrix} -\sigma & \sigma & 0 \\ r - Z(t) & -1 & -X(t) \\ Y(t) & X(t) & -b \end{pmatrix}\begin{pmatrix} \delta X \\ \delta Y \\ \delta Z \end{pmatrix} = \mathbf{L}(t; \mathbf{X}(0))\,\delta\mathbf{X}. \tag{4.3.51}$$

In the light of the dissipative nature of the dynamics, the sum of all three global Lyapunov exponents is negative. Indeed, because $Tr\left(\mathbf{L}(t;\mathbf{x}(0))\,\mathbf{P}^{(3)}(t)\right) = Tr(\mathbf{L}) = -(\sigma + 1 + b)$ is constant on the attractor,

$$\mu_1 + \mu_2 + \mu_3 = -(\sigma + 1 + b). \tag{4.3.52}$$

The sum of the first two global Lyapunov exponents is

$$\mu_1 + \mu_2 = \limsup_{t\to\infty}\ \sup_{\mathbf{X}(0)\in\mathscr{A}}\ \sup_{\mathbf{P}^{(2)}(0)} \left\{\frac{1}{t}\int_0^t Tr\left(\mathbf{L}^H\left(t', \mathbf{X}(0)\right)\mathbf{P}^{(2)}(t')\right)\,dt'\right\}, \tag{4.3.53}$$

and because some two-dimensional areas on the attractor do expand, these will be positive. An upper bound on the sum of the first two global Lyapunov exponents, i.e., an estimate for the largest growth rate of infinitesimal areas on the attractor, will lead to an upper bound on the Lyapunov dimension of the global attractor.

The area spanned by two infinitesimal vectors $\delta\mathbf{X}^{(1)}$ and $\delta\mathbf{X}^{(2)}$ is the cross (wedge) product $\delta\mathbf{S} = \delta\mathbf{X}^{(1)} \times \delta\mathbf{X}^{(2)}$, with direction normal to the

[1] In the convection interpretation, these dynamics correspond to the unstable quiescent state (the origin) developing a roll structure which chaotically reverses its sense of rotation.

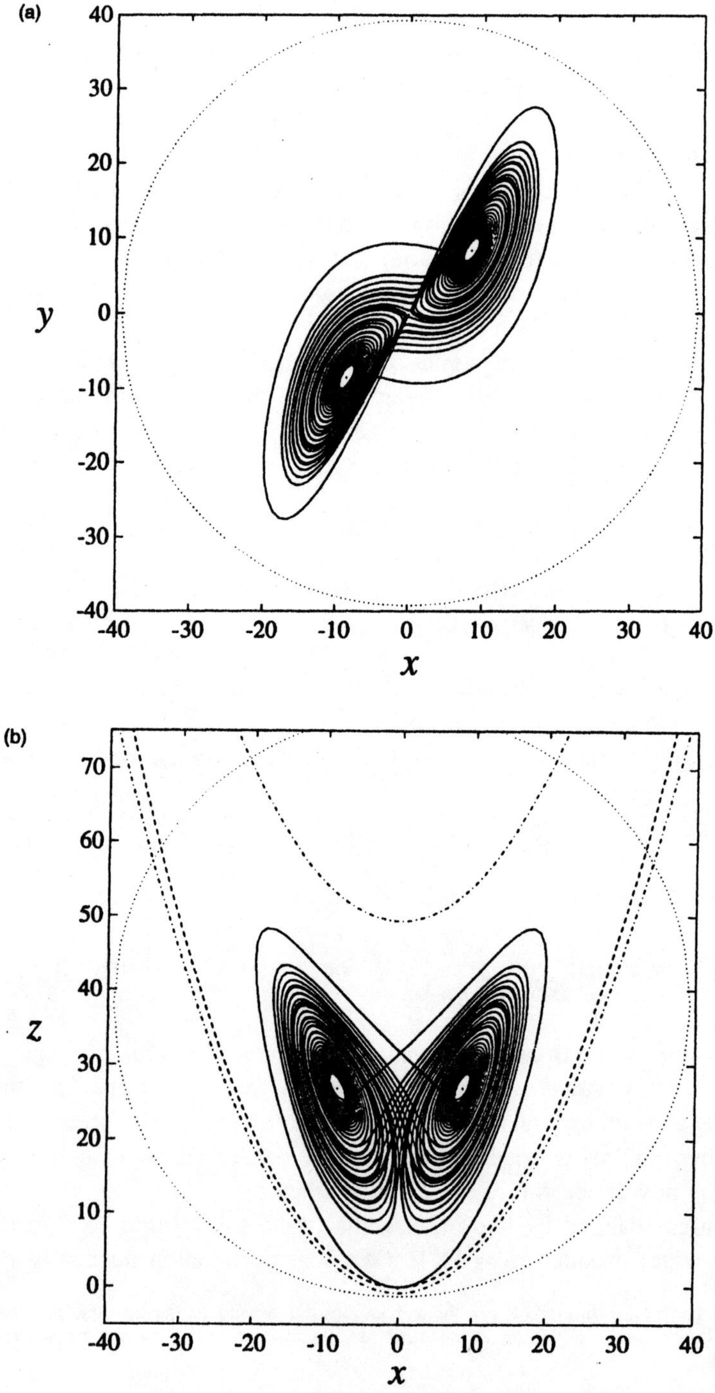

Fig. 4.4. (*a*) and (*b*) for caption see facing page

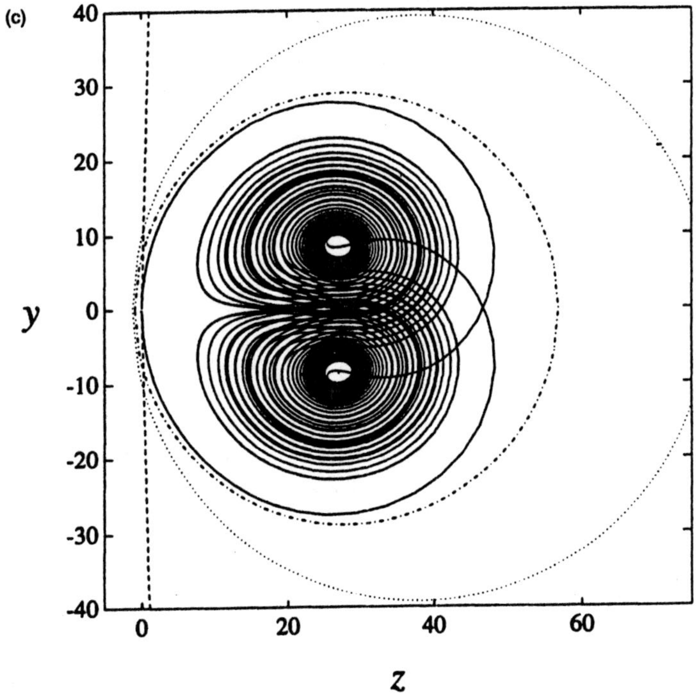

Fig. 4.4. (*a*) Projection of the Lorenz attractor onto the $x - y$ plane. The analytical bound given by the Lyapunov functional K is also shown (dotted line). (*b*) Projection of the Lorenz attractor onto the $x - z$ plane. Analytical bounds determined from K (dotted), I (dash-dot), and G (dashed) are also shown. (*c*) Projection of the Lorenz attractor onto the $z - y$ plane. Bounds determined from K dotted, J (dash-dot), and G (dashed) are shown.

area. Each of the edge vectors evolves according to

$$\frac{d\delta \mathbf{X}^{(i)}}{dt} = \mathbf{L}\delta \mathbf{X}^{(i)}, \tag{4.3.54}$$

so the area vector evolves according to

$$\frac{d\delta \mathbf{S}}{dt} = \left\{ Tr(\mathbf{L}) - \mathbf{L}^T \right\} \delta \mathbf{S}. \tag{4.3.55}$$

In the case of the Lorenz equations, then,

$$\frac{d}{dt}\begin{pmatrix} \delta S_X \\ \delta S_Y \\ \delta S_Z \end{pmatrix} = \begin{pmatrix} -(1+b) & Z(t)-r & -Y(t) \\ -\sigma & -(\sigma+b) & -X(t) \\ 0 & X(t) & -(\sigma+1) \end{pmatrix} \begin{pmatrix} \delta S_X \\ \delta S_Y \\ \delta S_Z \end{pmatrix}. \tag{4.3.56}$$

The expansion or contraction rate of areas are the real parts of the three eigenvalues of the matrix in (4.3.56). The range of (X, Y, Z) on the attractor is inside the sphere $K \leq K_{max}$, the solid cylinder $J \leq J_{max}$, and the ellipsoid $G \leq G_{max}$. The sum of the first two global Lyapunov exponents can then be bounded by computing the real parts of the eigenvalues (numerically, for example, for particular parameter values), and searching out the maximum over the intersection of the interiors of the ball, the cylinder and the ellipsoid. For the commonly used parameter values $\sigma = 10, b = 8/3$ and $r = 28$, this yields

$$\mu_1 + \mu_2 \leq 9.161\ldots \tag{4.3.57}$$

This maximum growth rate of areas on the attractor is realized at the origin $(X, Y, Z) = (0, 0, 0)$ which, as a fixed point, is contained in the global attractor. Hence the bound in (4.3.57) is optimal.

To estimate the dimension of the attractor, we use (4.2.19) with $\tilde{n} = 3$ and recall (4.3.52). For the parameter values $\sigma = 10, b = 8/3$ and $r = 28$, this gives the a priori bound on the (Lyapunov or Hausdorff) dimension of the global attractor for the Lorenz equations as

$$d_L = 2 + \frac{\mu_1 + \mu_2}{-\mu_3} = 2 + \frac{\mu_1 + \mu_2}{\sigma + 1 + b + \mu_1 + \mu_2} \leq 2.401\ldots \tag{4.3.58}$$

4.4 References and further reading

There are many texts on the modern theory of dynamical systems and chaos, for instance Drazin [9]. Attractors and their dimensions are discussed in a review article by Farmer, Ott, and Yorke [10]. The rigorous connection between the Lyapunov exponents and the attractor dimension was made by Constantin and Foias [11]. Lower bounds on the dimension of the attractor for the Navier-Stokes equations have been developed by Babin and Vishik [13]. More advanced reading can be found in the books by Constantin and Foias [12] and Temam [14] but mainly for PDEs. The Lorenz equations were introduced by Lorenz [15] and are the subject of the book by Sparrow [16]. The Lyapunov functionals I and J for the Lorenz equations were first introduced in an unpublished paper by Y. Treve [17]. Attractor dimension estimates for the Lorenz equations have been developed by Constantin, Foias, Manley, and Temam [18] and Eden, Foias, and Temam [19].

Exercises

1 The "middle third" Cantor set is constructed by removing the middle third of $[0, 1]$ and then removing the middle thirds of the remaining two intervals, etc. Compute its capacity dimension.

2 Consider the dynamical system

$$\dot{x} = (1 + y^2)x - x^3 \quad \text{(E4.1)}$$

$$\dot{y} = -(1 + y^2)y. \quad \text{(E4.2)}$$

Locate the fixed points and determine their stability. Prove that the global attractor is contained in the set $x^2 + y^2 \leq 1$. Show that the global attractor is actually the set $[-1, 1] \times \{0\}$.

3 Insert (4.3.9) and (4.3.10) into the Boussinesq equations and truncate the extra nonlinear terms to derive the Lorenz equations.

4 Verify equations (4.3.35), (4.3.38) and (4.3.49).

5 Solve the Lorenz equations numerically and, based on the definition of (4.2.27), compute the first Lyapunov exponent.

6 If X, Y and Z satisfy the Lorenz equations, show that the cone

$$\sigma(Y^2 + Z^2) = rX^2,$$

expels orbits to its exterior when $\sigma \geq 1$ and $\sigma \geq b$.

5

On the existence, uniqueness, and regularity of solutions

5.1 Introduction

In this chapter we explore the problem of establishing the existence and uniqueness of solutions of the Navier-Stokes equations. The existence issue touches on the question of the self-consistency of the physical model embodied in the Navier-Stokes equations; if no solutions exist, then the theory is empty. The question of uniqueness relates to the predictive power of the model. In classical mechanics, uniqueness of the solutions of the equations of motion is the cornerstone of classical determinism. A breakdown of uniqueness signals the introduction of other effects, effects which are not contained in the dynamical equations, into the system's evolution. For the incompressible Navier-Stokes equations these are not trivial questions either mathematically or physically.

In the next section we review the standard theory of existence and uniqueness of solutions of ordinary differential equations (ODEs), stressing the importance of either a global Lipschitz condition, or a local Lipschitz condition along with a priori bounds, for global and/or local existence and for uniqueness. The subsequent section is concerned with constructing Galerkin approximations and the so-called "weak" solutions of the Navier-Stokes partial differential equations (PDEs). Existence questions are necessarily more involved for PDEs due to the fact that there may be a selection of function spaces available in which to look for solutions, each of which typically admits a variety of topologies, and hence a variety of notions of convergence. Without entering into the details of the functional analysis (which is not the aim of this book and for which a number of complete, authoritative texts already exist) we set out to explain the notion of weak solutions and to focus on the essential ingredients used to prove their existence. In the last section, the question of uniqueness of the weak solutions is addressed. We discuss related regularity issues, along with the notion of "strong" or "classical" solutions of the PDE. It is here that the central physical distinction between the $2d$ and $3d$ Navier-Stokes equations manifests itself mathematically, leaving us with a fundamental – and to date, unsatisfactorily answered – question: Do the $3d$ Navier-Stokes equations constitute a classically deterministic dynamical model?

5.2 Existence and uniqueness for ODEs

We begin by reviewing and discussing the standard theory of existence and uniqueness for systems of ODEs. Consider the N variable (without loss of generality, autonomous) system

$$\frac{d\mathbf{z}}{dt} = \mathbf{F}(\mathbf{z}), \tag{5.2.1}$$

where $\mathbf{z} \in \mathbf{C}^N$ and the "forcing" function $\mathbf{F} : \mathbf{C}^N \to \mathbf{C}^N$, along with the initial condition

$$\mathbf{z}(0) = \mathbf{z}_0 \in \mathbf{C}^N. \tag{5.2.2}$$

The analysis proceeds by first constructing a solution $\mathbf{z}(t)$, and then establishing its uniqueness among a reasonable class of functions.

A sufficient condition for the existence and uniqueness of a solution is a *Lipschitz condition* on $\mathbf{F}$. That is, for the moment, we assume that the forcing function admits a constant $K < \infty$ such that for all $\mathbf{z}, \mathbf{z}' \in \mathbf{C}^N$,

$$|\mathbf{F}(\mathbf{z}) - \mathbf{F}(\mathbf{z}')| \leq K|\mathbf{z} - \mathbf{z}'|. \tag{5.2.3}$$

An acceptable K is referred to as a *Lipschitz constant* for $\mathbf{F}$.

Then a solution to the integral formulation of the ODE, written

$$\mathbf{z}(t) = \mathbf{z}_0 + \int_0^t \mathbf{F}(\mathbf{z}(t'))dt', \tag{5.2.4}$$

may be constructed by *Picard iteration.* Starting from $n = 0$, define the functions $\mathbf{z}_n(t)$ iteratively according to

$$\mathbf{z}_{n+1}(t) = \mathbf{z}_0 + \int_0^t \mathbf{F}(\mathbf{z}_n(t'))dt'. \tag{5.2.5}$$

For times t in an interval $[0, T]$, where $T < K^{-1}$, this sequence of functions converges in the Banach space $C([0, T]; \mathbf{C}^N)$ of $\mathbf{C}^N$-valued continuous functions on $[0, T]$ with the norm

$$\|\mathbf{z}(\cdot)\| = \sup_{0 \leq t \leq T} |\mathbf{z}(t)|. \tag{5.2.6}$$

Convergence of a sequence in $C([0,T];\mathbf{C}^N)$ means uniform pointwise convergence on $[0,T]$.

This is established by considering the difference between successive iterates and estimating as follows:

$$\begin{aligned}
\|\mathbf{z}_{n+1}-\mathbf{z}_n\| &= \sup_{0\le t\le T}\left|\int_0^t \{\mathbf{F}(\mathbf{z}_n(t'))-\mathbf{F}(\mathbf{z}_{n-1}(t'))\}\,dt'\right| \\
&\le \sup_{0\le t\le T}\int_0^t K|\mathbf{z}_n(t')-\mathbf{z}_{n-1}(t')|\,dt' \\
&\le KT\sup_{0\le t\le T}|\mathbf{z}_n(t)-\mathbf{z}_{n-1}(t)| \\
&= KT\|\mathbf{z}_n-\mathbf{z}_{n-1}\|, \qquad (5.2.7)
\end{aligned}$$

where $KT<1$. Hence the mapping $\mathbf{z}_n \to \mathbf{z}_{n+1}$ is a contraction map in the complete space $C([0,T];\mathbf{C}^N)$, and so the sequence $\mathbf{z}_n$ is a Cauchy sequence with a unique limit there (see exercise 5.1). Call the limit $\mathbf{z}(t)$. It is a candidate for the solution of (5.2.4).

To show that the $\mathbf{z}(t)$ so constructed is indeed a solution, we must verify that it satisfies the integral formulation of the ODE in (5.2.4). That this is so follows from the fact that $\mathbf{z}_n \to \mathbf{z}$ in $C([0,T];\mathbf{C}^N)$, along with the continuity of $\mathbf{F}$ implies $\mathbf{F}(\mathbf{z}_n) \to \mathbf{F}(\mathbf{z})$ in $C([0,T];\mathbf{C}^N)$. Then the right-hand side of (5.2.5) also converges uniformly on $[0,T]$ to the right-hand side of (5.2.4). Hence the limit function $\mathbf{z}(t)$ solves the integral form of the differential equation. And because $\mathbf{F}(\mathbf{z}(t))$ is continuous on $[0,T]$, the integral may be differentiated so that $\mathbf{z}(t)$ is differentiable and actually satisfies the ODE (it clearly satisfies the initial condition).

The solution constructed via Picard iteration is the unique continuous solution on $[0,T]$ starting from the given initial condition. To see this, assume that two continuous functions $\mathbf{z}(t)$ and $\mathbf{z}'(t)$ satisfy the ODE (5.2.1) and the initial condition (5.2.2). Then the difference, $\mathbf{y}(t)=\mathbf{z}'(t)-\mathbf{z}(t)$, satisfies

$$\frac{d\mathbf{y}}{dt}=\mathbf{F}(\mathbf{z}+\mathbf{y})-\mathbf{F}(\mathbf{z}) \qquad (5.2.8)$$

with the initial condition

$$\mathbf{y}(0)=0. \qquad (5.2.9)$$

Using the Lipschitz condition on $\mathbf{F}$ we deduce

$$\begin{aligned}
\frac{d}{dt}|\mathbf{y}|^2 &= \mathbf{y}^*\{\mathbf{F}(\mathbf{z}+\mathbf{y})-\mathbf{F}(\mathbf{z})\}+\mathbf{y}\cdot\{\mathbf{F}(\mathbf{z}+\mathbf{y})^*-\mathbf{F}(\mathbf{z})^*\} \\
&\le 2K|\mathbf{y}|^2. \qquad (5.2.10)
\end{aligned}$$

This differential inequality may be integrated via Gronwall's inequality (Lemma 2.1), yielding

$$|\mathbf{y}(t)|^2 \le |\mathbf{y}(0)|^2 \exp\{2Kt\} = 0 \times \exp\{2Kt\} = 0. \tag{5.2.11}$$

Hence any two solutions starting from the same initial condition are the same, and uniqueness is established.

We have thus constructed the *local* unique solution of the ODE, qualified as such because it is only valid locally in time, for times $t \le T$. The local solution may be continued to the *global* solution, valid for all $t > 0$. The standard argument proceeds by noting that the norm of the solution is bounded in $C([0,T];\mathbf{C}^N)$:

$$\begin{aligned} \|\mathbf{z}\| &= \lim_{n\to\infty} \|\mathbf{z}_n\| \\ &\le \lim_{n\to\infty} \left\{ \sum_{j=1}^{n} \|\mathbf{z}_j - \mathbf{z}_{j-1}\| + |\mathbf{z}_0| \right\} \\ &\le \lim_{n\to\infty} \left\{ \sum_{j=0}^{n} (KT)^j |\mathbf{F}(\mathbf{z}_0)|\, T + |\mathbf{z}_0| \right\} \\ &= \frac{|\mathbf{F}(\mathbf{z}_0)|T}{1-KT} + |\mathbf{z}_0|, \end{aligned} \tag{5.2.12}$$

where we have utilized the explicit bound, deduced via induction from (5.2.7), that $\|\mathbf{z}_{n+1} - \mathbf{z}_n\| \le (KT)^n |\mathbf{F}(\mathbf{z}_0)| T$ for $n \ge 0$. Then we know that $\mathbf{z}(T)$ is finite, and so the Picard iteration may be started over at time T using $\mathbf{z}(T)$ as the initial condition to construct the unique continuous solution up until time $2T$. This may be repeated over and over, explicitly constructing the unique continuous solution over successive intervals of length T, resulting in the unique global continuous solution of the ODE.

A simple example of an ODE satisfying a Lipschitz condition is the one dimensional ($N = 1$) linear problem

$$\frac{dz}{dt} = \alpha z, \tag{5.2.13}$$

where the parameter α is a constant. The smallest Lipschitz constant is $K = |\alpha|$, and the Picard iterates are

$$\begin{aligned} z_0(t) &= z(0), \\ z_1(t) &= z(0) + z(0)\alpha t, \\ z_2(t) &= z(0) + z(0)\alpha t + z(0)\frac{1}{2}\alpha t^2, \end{aligned} \tag{5.2.14}$$

etc., which are easily seen to converge uniformly on bounded time intervals (and in particular on $[0, \alpha^{-1}]$) to the solution

$$z(t) = z(0) \exp\{\alpha t\}. \tag{5.2.15}$$

The theory of existence and uniqueness of solutions of ODEs satisfying a Lipschitz condition is thus quite complete and, as outlined above, uses only elementary analysis. The Lipschitz condition is very restrictive, however, and in applications one often encounters ODEs which do not satisfy such a strong condition. An elementary example is the one-variable nonlinear ODE

$$\frac{dz}{dt} = \alpha z^3. \tag{5.2.16}$$

The analysis becomes more delicate in the absence of a Lipschitz constant, and the distinction between local and global solutions becomes more than just a formality. What makes things work at all for ODEs like (5.2.16) is the fact that although the right hand of (5.2.16) doesn't satisfy a Lipschitz condition as defined above, it is a "smooth" function which does satisfy such a condition over restricted ranges of the variable.

So consider a system of ODEs as in (5.2.1), but for which no finite constant can be found to play the role of K in the Lipschitz condition in (5.2.3), and restrict attention to the case where $\mathbf{F}(\mathbf{z}_0) \neq 0$ (for otherwise a solution is simply the constant, $\mathbf{z}(t) = \mathbf{z}_0$). Then we cannot rely on the previous proof that the Picard iterates converge, and we are at a loss as to how to construct a solution. In all generality this could be a real problem. In practice, however, we can extend the construction and the local existence and uniqueness proofs to ODEs that are "locally" Lipschitz. We say that $\mathbf{F} : \mathbf{C}^N \to \mathbf{C}^N$ satisfies a *local Lipschitz condition* if there is a $K(\mathbf{z}, \mathbf{z}') < \infty$ such that,

$$|\mathbf{F}(\mathbf{z}) - \mathbf{F}(\mathbf{z}')| \leq K(\mathbf{z}, \mathbf{z}')|\mathbf{z} - \mathbf{z}'|, \tag{5.2.17}$$

and the "local" Lipschitz constant $K(\mathbf{z}, \mathbf{z}')$ is uniformly bounded on bounded sets in $\mathbf{C}^N \times \mathbf{C}^N$.

The strategy for dealing with this kind of problem is to observe that in a neighborhood of the initial condition, the local Lipschitz condition implies the usual Lipschitz condition (with a Lipschitz constant which generally increases with the size of the neighborhood). Then as long as the candidate solution stays in that neighborhood – which it ought to do for at least a short time – the existence and uniqueness proof given above will apply. So, given the initial condition $\mathbf{z}(0) = \mathbf{z}_0$, choose a positive number $c > |\mathbf{z}_0|$. Let

$$K' = \sup_{|\mathbf{z}|,|\mathbf{z}'|<2c} K(\mathbf{z}, \mathbf{z}'), \tag{5.2.18}$$

which is the effective Lipschitz constant, and let

$$T = \frac{1}{\frac{|\mathbf{F}(\mathbf{z}_0)|}{c} + K'}. \tag{5.2.19}$$

We can now find the unique solution in $C([0,T];\mathbf{C}^N)$. Defining the Picard iterates $\mathbf{z}_n(t)$ as in (5.2.5), we note that if we can show that successive iterates stay within the ball of radius $2c$ in $C([0,T];\mathbf{C}^N)$, then K' remains an effective Lipschitz constant and the estimates in (5.2.7) follow (with K' in place of K). We may then conclude that the iterates will converge.

A simple induction argument shows that the successive iterates are bounded by $2c$. First, for $n = 1$,

$$\mathbf{z}_1(t) = \mathbf{z}_0 + \mathbf{F}(\mathbf{z}_0)t, \tag{5.2.20}$$

so

$$\begin{aligned} |\mathbf{z}_1| &\le |\mathbf{z}_0| + |\mathbf{F}(\mathbf{z}_0)|T \\ &\le |\mathbf{z}_0| + c \\ &< 2c. \end{aligned} \tag{5.2.21}$$

Now assume that each of $\|\mathbf{z}_1\|, \|\mathbf{z}_2\|, \ldots, \|\mathbf{z}_n\| < 2c$. Then

$$\begin{aligned} \|\mathbf{z}_{n+1} - \mathbf{z}_n\| &= \sup_{0\le t\le T} \left| \int_0^t \{\mathbf{F}(\mathbf{z}_n(t')) - \mathbf{F}(\mathbf{z}_{n-1}(t'))\}\, dt' \right| \\ &\le \sup_{0\le t\le T} \int_0^t K'|\mathbf{z}_n(t') - \mathbf{z}_{n-1}(t')|dt' \\ &\le K'T\|\mathbf{z}_n - \mathbf{z}_{n-1}\|, \end{aligned} \tag{5.2.22}$$

so that for values of m from 1 up to n,

$$\|\mathbf{z}_{m+1} - \mathbf{z}_m\| \le (K'T)^m \|\mathbf{z}_1 - \mathbf{z}_0\| = (K'T)^m |\mathbf{F}(\mathbf{z}_0)|T. \tag{5.2.23}$$

Hence,

$$\begin{aligned} \|\mathbf{z}_{n+1}\| &\le \|\mathbf{z}_{n+1} - \mathbf{z}_n\| + \|\mathbf{z}_n - \mathbf{z}_{n-1}\| + \ldots + \|\mathbf{z}_1 - \mathbf{z}_0\| + \|\mathbf{z}_0\| \\ &\le \left\{(K'T)^n + (K'T)^{n-1} + \ldots + 1\right\} |\mathbf{F}(\mathbf{z}_0)|T + \|\mathbf{z}_0\| \\ &< \frac{|\mathbf{F}(\mathbf{z}_0)|T}{1 - K'T} + \|\mathbf{z}_0\| \\ &= c + \|\mathbf{z}_0\| \\ &< 2c, \end{aligned} \tag{5.2.24}$$

concluding the induction and establishing uniform convergence of the iterates.

Uniform convergence of the Picard iterates then leads, as before, to the conclusion that the limit solves the ODE uniquely up to time T and provides a local solution of the problem. Note that although we have some freedom in the choice of T, it cannot generally be taken arbitrarily large. We have a lot of freedom choosing c, but a smaller choice for c minimizes K' at the cost of increasing the $|\mathbf{F}(\mathbf{z}_0)|/c$ term in the denominator of (5.2.19), while ever larger choices for c eventually increase K', thereby also reducing T. Ideally, the local solution can be elevated to a global solution, but by attempting to mimic the extension used in the Lipschitz case we can see how the process may not succeed.

The bound on the iterates in (5.2.24) leads to an explicit bound on the solution up to and at time T, namely

$$|\mathbf{z}(T)| \leq 2c. \tag{5.2.25}$$

The iteration procedure may be repeated starting from time $t = T$ with the initial condition $z(t)$, by choosing a new $c' > 2c \geq |\mathbf{z}(T)|$, and computing a new K'' based on c'. This will work, allowing us to extend the solution uniquely over an interval of length T' (basing its computation on K'', c', and an estimate of $|\mathbf{F}(\mathbf{z}(T))|$) up to time $T + T'$. The process may then be repeated for another interval of time, say, T''. The hitch is that generally the successive intervals $T', T'', \ldots$ could become shorter and shorter as the norm of the solution grows and the local Lipschitz constant increases. If the sequence $T', T'', \ldots$ decreases so fast as to be summable, then the solution may not be continued to arbitrarily large times.

The example in (5.2.16) is a case in point. Taking z and α real for simplicity, the exact solution starting from $z(0) = z_0$ is

$$z(t) = \frac{z_0}{\sqrt{1 - 2z_0^2 \alpha t}}. \tag{5.2.26}$$

If $\alpha > 0$ and $z_0 \neq 0$, then the norm of the solution diverges at time $T^* = (2z_0^2\alpha)^{-1}$. Any attempt to construct a continuous solution for arbitrarily long times, at least for general initial conditions and parameter values, is doomed to failure. The construction procedure outlined above for the locally Lipschitz ODEs exactly captures this feature: Note that a choice of $c \sim |z_0|$ in (5.2.18) and (5.2.19) gives a time $T \sim (z_0^2|\alpha|)^{-1}$ which has the same scaling with respect to the z_0 and α as T^*.

This does not mean that unique, bounded, continuous solutions cannot be found for non-Lipschitz ODEs, only that more information is necessary and that details of parameter and initial condition values generally come into play. This is where the role of a priori bounds on the solutions becomes crucial. Even in the absence of a global Lipschitz condition, if the norm of local solutions can be controlled sufficiently by some other means, so that the local Lipschitz condition remains effective, then the construction procedure can be extended to arbitrarily long times. The example ODE in (5.2.16) illustrates this idea. If $\alpha < 0$, the exact solution in (5.2.26) is bounded and continuous for all positive times. The question is how this might have been deduced in the absence of an explicit solution.

Consider the ODE in (5.2.16) with $\alpha < 0$ (and z real). Because it satisfies a local Lipschitz condition, a unique continuous solution exists up until a time T as given in (5.2.19). The bound (5.2.24) of the norm on the solution at time T deduced from the Picard iteration procedure is then an overestimate which may be improved as follows. The differential equation satisfied by z^2 is

$$\frac{d}{dt}z^2 = -2|\alpha|z^4, \tag{5.2.27}$$

implying the differential inequality

$$\frac{d}{dt}z^2 \leq 0. \tag{5.2.28}$$

This means that the magnitude of z decreases as t increases, at least as long as the solution exists. Integrating the differential inequality in (5.2.28), we conclude that

$$|z(T)| \leq |z_0|. \tag{5.2.29}$$

So the norm at time T is in fact less than or equal to the norm of the initial condition and the iteration procedure may then be repeated for another interval of time of length T (the *same* interval T) to extend the unique continuous solution up to time $2T$. Appealing again to the evolution of the magnitude in (5.2.27), we deduce that $|z(2T)| \leq |z(T)| \leq |z(0)|$ and so the process can be repeated for another interval of the same length T. Thus the a priori bound deduced from (5.2.27) leads to sufficient control to allow for the unique global continuous solution to exist, and – just as important from a practical standpoint – to be shown to exist.

Without a local Lipschitz condition (or something akin to it) local uniqueness of solutions may be a problem. The following elementary example illustrates what can happen. Consider the ODE

$$\frac{dz}{dt} = \sqrt{|z|} \tag{5.2.30}$$

with the initial condition

$$z(0) = 0. \tag{5.2.31}$$

The function $\sqrt{|z|}$ is not locally Lipschitz in a neighborhood of $z(0) = 0$, and this manifests itself as a loss of uniqueness of (continuous) solutions. Indeed, both

$$z(t) = 0 \tag{5.2.32}$$

and

$$z(t) = \frac{1}{4}t^2 \tag{5.2.33}$$

are continuous solutions satisfying both the differential equation and the initial condition.

The lessons to be learned are fourfold:

1 A global Lipschitz condition leads to global existence and uniqueness, whereas
2 a local Lipschitz condition generally leads only to local existence and uniqueness. However,
3 a local Lipschitz condition supplemented with some additional a priori bounds, if appropriate bounds exist and can be found, leads to global existence and uniqueness. And finally,
4 without a local Lipschitz condition our construction of a continuous solution via Picard iteration is suspect and moreover, even if a continuous solution can be found by some other means, it is not necessarily unique.

5.3 Galerkin approximations and weak solutions of the Navier-Stokes equations

In the previous section, solutions of ODEs were constructed by writing down explicit successive approximations (the Picard iterates) and then showing that the approximations converged, in a specific sense, to exact solutions. The Navier-Stokes equations are PDEs which can be considered as an infinite system of ODEs. To manufacture global solutions, the idea is to write down explicit successive approximations consisting of solutions of finite systems of ODEs (the Galerkin approximates) and then to show that the approximations converge, in a certain sense, to solutions of the PDEs. The ODEs will be treated with the methods of the previous section: It behooves us to show that they satisfy a certain Lipschitz condition, so that local solutions can be constructed, and then to provide appropriate a priori estimates allowing the local solutions to be extended to global solutions. The solutions of the PDEs produced in this way are known as "weak" solutions because the original equation can only be shown to be satisfied in a weak (smeared) sense.

For simplicity and definiteness, we consider the Navier-Stokes equations with periodic boundary conditions on $\Omega = [0, L]^d$ in the presence of a body force which is constant in time:

$$\frac{\partial \mathbf{u}}{\partial t} + \mathbf{u} \cdot \nabla \mathbf{u} + \nabla p = \nu \Delta \mathbf{u} + \mathbf{f}(\mathbf{x}), \tag{5.3.1}$$

$$\nabla \cdot \mathbf{u} = 0, \tag{5.3.2}$$

with initial condition

$$\mathbf{u}(\mathbf{x}, 0) = \mathbf{u}_0(\mathbf{x}). \tag{5.3.3}$$

The initial condition $\mathbf{u}_0(\mathbf{x})$ is a periodic, square integrable, divergence-free vector field. Moreover, we consider the case of a mean zero $\mathbf{f}(\mathbf{x})$ and mean zero $\mathbf{u}_0(\mathbf{x})$; the evolution preserves this mean zero property for $\mathbf{u}(\mathbf{x}, t)$. This is necessary if we are going to use Poincaré's inequality with periodic boundary conditions. Mass units have been chosen so that the density has been set to unity. The problem in the presence of rigid boundaries is significantly more involved, and we refer the reader to the references for results in this area.

As discussed in Chapter 3, a velocity field may be Fourier transformed according to

$$\hat{\mathbf{u}}(\mathbf{k}, t) = \int_\Omega \exp\{-i\mathbf{k} \cdot \mathbf{x}\} \mathbf{u}(\mathbf{x}, t)\, d^d x, \tag{5.3.4}$$

with inverse transform

$$\mathbf{u}(\mathbf{x}, t) = L^{-d} \sum_{\mathbf{k}} \exp\{i\mathbf{k} \cdot \mathbf{x}\} \hat{\mathbf{u}}(\mathbf{k}, t), \tag{5.3.5}$$

where the discrete wavenumbers $\mathbf{k}$ are

$$\mathbf{k} = \sum_{j=1}^{d} \mathbf{e}_j \frac{2\pi n_j}{L}, \qquad n_j = 0, \pm 1, \pm 2, \ldots, \tag{5.3.6}$$

with $\mathbf{e}_j$ the unit vector in the jth direction. Because the velocity vector field is real, the complex Fourier coefficients and their complex conjugates are related by

$$\hat{\mathbf{u}}^*(\mathbf{k},t) = \hat{\mathbf{u}}(-\mathbf{k},t). \tag{5.3.7}$$

The time evolution of the wavenumber components are an infinite set of ODEs computed as the Fourier transform of the Navier-Stokes equations (recall 3.3.9):

$$\frac{d}{dt}\hat{\mathbf{u}}(\mathbf{k},t) = -\nu k^2\hat{\mathbf{u}}(\mathbf{k},t) + \frac{i}{L^3}\left(\mathbf{I} - \frac{\mathbf{kk}}{k^2}\right)\cdot \sum_{\mathbf{k}'+\mathbf{k}''=\mathbf{k}} \hat{\mathbf{u}}(\mathbf{k}',t)\cdot\mathbf{k}''\hat{\mathbf{u}}(\mathbf{k}'',t) + \hat{\mathbf{f}}(\mathbf{k}), \tag{5.3.8}$$

$$\mathbf{k}\cdot\hat{\mathbf{u}}(\mathbf{k},t) = 0, \tag{5.3.9}$$

where $\mathbf{I}$ is the unit tensor, $\mathbf{I} - \mathbf{kk}/k^2$ is the projector onto divergence-free vector fields in wavenumber space, and $\hat{\mathbf{f}}(\mathbf{k})$ is the Fourier transform of the body force (without loss of generality, $\mathbf{k}\cdot\hat{\mathbf{f}}(\mathbf{k}) = 0$). This is an infinite set of ODEs because the sum over $\mathbf{k}'$ and $\mathbf{k}''$ couples all the dynamical variables. The initial conditions for the Fourier transformed variables are

$$\hat{\mathbf{u}}_0(\mathbf{k}) = \int_\Omega \exp\{-i\mathbf{k}\cdot\mathbf{x}\}\mathbf{u}_0(\mathbf{x})\, d^dx. \tag{5.3.10}$$

Note that without loss of generality, we may presume that $\hat{\mathbf{u}}_0(0,t) = 0$ for all $t \geq 0$.

The Galerkin approximations are truncated Fourier expansions. In particular, for a given integer $N > 0$, we consider the finite collection of complex variables $\hat{\mathbf{u}}^N(\mathbf{k},t)$ for

$$\mathbf{k} = \sum_{j=1}^{d} \mathbf{e}_j \frac{2\pi n_j}{L}, \qquad n_j = \pm 1, \pm 2, \ldots, \pm N, \tag{5.3.11}$$

where the values of $\hat{\mathbf{u}}^N$ satisfy the reality condition

$$\hat{\mathbf{u}}^N(-\mathbf{k},t) = \hat{\mathbf{u}}^N(\mathbf{k},t)^*. \tag{5.3.12}$$

We are working with the case $\hat{\mathbf{u}}^N(0,t) = 0$. The finite system of coupled ODEs for the $\hat{\mathbf{u}}^N(\mathbf{k},t)$s is

$$\begin{aligned}\frac{d}{dt}\hat{\mathbf{u}}^N(\mathbf{k},t) &= -\nu k^2\hat{\mathbf{u}}^N(\mathbf{k},t) \\ &+ \frac{i}{L^3}\left(\mathbf{I} - \frac{\mathbf{kk}}{k^2}\right)\cdot \sum_{\mathbf{k}'+\mathbf{k}''=\mathbf{k}} \left[\hat{\mathbf{u}}^N(\mathbf{k}',t)\cdot\mathbf{k}''\hat{\mathbf{u}}^N(\mathbf{k}'',t)\right] + \hat{\mathbf{f}}(\mathbf{k}),\end{aligned} \tag{5.3.13}$$

$$\mathbf{k}\cdot\hat{\mathbf{u}}^N(\mathbf{k},t) = 0, \tag{5.3.14}$$

where the sum over $\mathbf{k}'$ extends over the range where both $\mathbf{k}'$ and $\mathbf{k}''$ have components between $-2\pi N/L$ and $2\pi N/L$, as in (5.3.11). The initial conditions are the truncation of the initial condition for the full problem:

$$\hat{\mathbf{u}}_0^N(\mathbf{k}) = \hat{\mathbf{u}}_0(\mathbf{k}), \qquad \text{for} \qquad \mathbf{k} = \sum_{j=1}^{d} \mathbf{e}_j \frac{2\pi n_j}{L}, \qquad n_j = \pm 1, \pm 2, \ldots, \pm N. \tag{5.3.15}$$

Hence, the Galerkin approximation's evolution looks like the transformed Navier-Stokes equations, but modes with components larger than $2\pi N/L$ are set to zero. Note that both the divergence-free condition in (5.3.14) and the relations in (5.3.12) are preserved by the evolution in (5.3.13), so that as long as the initial condition satisfies $\mathbf{k} \cdot \hat{\mathbf{u}}^N(\mathbf{k}, 0) = 0$ and $\hat{\mathbf{u}}^N(\mathbf{k}, 0) = \hat{\mathbf{u}}^N(-\mathbf{k}, 0)^*$, then $\mathbf{k} \cdot \hat{\mathbf{u}}^N(\mathbf{k}, t) = 0$ and $\hat{\mathbf{u}}^N(\mathbf{k}, t) = \hat{\mathbf{u}}^N(-\mathbf{k}, t)^*$ for all $t > 0$. Net the constraints in (5.3.14), the system (5.3.13) is really $(d-1)\left(N^d + dN\right)$ coupled ODEs for the complex components of $\hat{\mathbf{u}}^N(\mathbf{k}, t)$.

Local existence of continuous solutions for $\hat{\mathbf{u}}^N(\mathbf{k}, t)$ is established by verifying the local Lipschitz condition satisfied by the right-hand side of (5.3.13). Indeed, when the divergence-free constraint in (5.3.14) is taken into account, these differential equations are of the form

$$\frac{dz_\alpha}{dt} = A_{\alpha\beta} z_\beta + B_{\alpha\beta\gamma} z_\beta z_\gamma = F_\alpha(\mathbf{z}), \tag{5.3.16}$$

for $M = (d-1)\left(N^d + dN\right)$ variables $\mathbf{z} = (z_1, z_2, \ldots, z_M)$, where $\alpha, \beta, \gamma = 1, \ldots, M$. Here $A_{\alpha\beta}$ and $B_{\alpha\beta\gamma}$ are arrays of bounded, complex constants involving only ν and L. The most brutal estimates yield the local Lipschitz condition

$$|\mathbf{F}(\mathbf{z}) - \mathbf{F}(\mathbf{z}')| \le \left\{ \max_{\alpha\beta} |A_{\alpha\beta}| + \max_{\alpha\beta\gamma} |B_{\alpha\beta\gamma}| \left(|\mathbf{z}| + |\mathbf{z}'|\right) \right\} |\mathbf{z} - \mathbf{z}'|. \tag{5.3.17}$$

Hence there exists a time $T > 0$, depending on the magnitude of the A and B arrays (which depend on ν, L and N) as well as the magnitude of the initial condition, such that there is a unique continuous solution $\hat{\mathbf{u}}_0^N(\mathbf{k}, t)$ on $[0, T]$.

Global existence for $\hat{\mathbf{u}}_0^N(\mathbf{k}, t)$ then follows from the observation that $\hat{\mathbf{u}}_0^N(\mathbf{k}, T)$ is bounded uniformly in T. This fact follows from the energy equation for the Galerkin approximations obtained from (5.3.13) by dotting with $\hat{\mathbf{u}}_0^N(\mathbf{k}, t)^*$, summing over $\mathbf{k}$, and adding the complex conjugate:

$$\frac{1}{2}\frac{d}{dt} \sum_{\mathbf{k}} |\hat{\mathbf{u}}^N(\mathbf{k}, t)|^2 = -\nu \sum_{\mathbf{k}} k^2 |\hat{\mathbf{u}}^N(\mathbf{k}, t)|^2 + \Re \sum_{\mathbf{k}} \hat{\mathbf{u}}^N(\mathbf{k}, t)^* \cdot \hat{\mathbf{f}}(\mathbf{k}). \tag{5.3.18}$$

This is the analogy to the energy evolution equation derived in the continuum in Chapter 2. The crucial point here is that the contributions from the nonlinear terms all cancel for the Galerkin approximations just as they do for the full equation. The nonlinear terms transfer energy (in the form of Fourier coefficient amplitude) between the modes, but they neither create nor destroy energy.

Noting that $k^2 \geq (2\pi/L)^2$ in the first sum on the right-hand side of (5.3.18) above, and using a Schwarz inequality on the second term, we have

$$\frac{1}{2}\frac{d}{dt}\sum_{\mathbf{k}} |\hat{\mathbf{u}}^N(\mathbf{k},t)|^2 \leq -\nu\left(\frac{2\pi}{L}\right)^2 \sum_{\mathbf{k}} |\hat{\mathbf{u}}^N(\mathbf{k},t)|^2 + \sqrt{\sum_{\mathbf{k}} |\hat{\mathbf{u}}^N(\mathbf{k},t)|^2}\sqrt{\sum_{\mathbf{k}} |\hat{\mathbf{f}}(\mathbf{k})|^2}. \tag{5.3.19}$$

Denoting

$$h(t) = \sqrt{\sum_{\mathbf{k}} |\hat{\mathbf{u}}^N(\mathbf{k},t)|^2}, \tag{5.3.20}$$

$$a = \nu\left(\frac{2\pi}{L}\right)^2, \tag{5.3.21}$$

and

$$f = \sqrt{\sum_{\mathbf{k}} |\hat{\mathbf{f}}(\mathbf{k})|^2}, \tag{5.3.22}$$

equation (5.3.19) reads

$$\frac{dh}{dt} \leq -ah + f. \tag{5.3.23}$$

Gronwall's lemma (Lemma 2.1) implies

$$h(t) \leq h(0)\exp(-at) + a^{-1}\left[1 - \exp(-at)\right] f, \tag{5.3.24}$$

so in terms of the original quantities,

$$\sum_{\mathbf{k}} |\hat{\mathbf{u}}^N(\mathbf{k},T)|^2 \leq \max\left\{\sum_{\mathbf{k}} |\hat{\mathbf{u}}^N(\mathbf{k},0)|^2, \frac{L^2}{4\pi^2\nu}\sum_{\mathbf{k}} |\hat{\mathbf{f}}(\mathbf{k})|^2\right\}. \tag{5.3.25}$$

Therefore unique continuous solution can be extended over another interval of time $T' > 0$, which itself only depends on $h(0)$ and f/a, at which time the same argument can be applied to bound $h(T + T')$ again by the same quantity. This process can be repeated indefinitely, and global existence and uniqueness of continuous solutions are established.

The continuum velocity vector fields reconstructed from the solutions of these ODEs,

$$\mathbf{u}^N(\mathbf{x},t) := L^{-d} \sum_{|k_j| \le \frac{2\pi N}{L}} \exp\{i\mathbf{k}\cdot\mathbf{x}\}\hat{\mathbf{u}}^N(\mathbf{k},t), \tag{5.3.26}$$

satisfy the PDE

$$\frac{\partial \mathbf{u}^N}{\partial t} + \mathbf{P}^N\left\{\mathbf{u}^N\cdot\nabla\mathbf{u}^N + \nabla p^N\right\} = \nu\Delta\mathbf{u}^N + \mathbf{P}^N\left\{\mathbf{f}(\mathbf{x})\right\}, \tag{5.3.27}$$

$$\nabla\cdot\mathbf{u}^N = 0, \tag{5.3.28}$$

with initial condition

$$\mathbf{u}^N(\mathbf{x},0) = \mathbf{P}^N\left\{\mathbf{u}_0(\mathbf{x})\right\}, \tag{5.3.29}$$

where $\mathbf{P}^N\{\cdot\}$ is the projection operator onto the first N Fourier modes. Its action on a periodic function $f(\mathbf{x})$ is explicitly

$$\mathbf{P}^N\{f\}(\mathbf{x}) = \int_\Omega G^N(\mathbf{x}-\mathbf{x}')f(\mathbf{x}')\,d^dx, \tag{5.3.30}$$

where $G^N(\mathbf{x})$ is the Dirichlet kernel

$$G^N(\mathbf{x}) = L^{-d} \sum_{|k_j| \le \frac{2\pi N}{L}} \exp\{i\mathbf{k}\cdot\mathbf{x}\}. \tag{5.3.31}$$

The projection operator is equivalent to the identity operator when acting on the Galerkin approximation:

$$\mathbf{P}^N\left\{\mathbf{u}^N(\mathbf{x},t)\right\} = \mathbf{u}^N(\mathbf{x},t). \tag{5.3.32}$$

The main idea is that the Galerkin equations (5.3.27)-(5.3.29) are nearly the projection of the full equations, but not quite. The difference is that the projection of the nonlinear term in the full equations, $\mathbf{P}^N\{\mathbf{u}\cdot\nabla\mathbf{u}+\nabla p\}$, is replaced by $\mathbf{P}^N\left\{\mathbf{P}^N\{\mathbf{u}\}\cdot\nabla\mathbf{P}^N\{\mathbf{u}\}+\nabla p\right\}$ in the Galerkin approximation in order to obtain a finite, closed set of equations for $\mathbf{u}^N$, which is intended to approximate $\mathbf{P}^N\{\mathbf{u}\}$.

The linear projection operator $\mathbf{P}^N$ has a number of important properties. First, like any projection operator, it satisfies

$$\left(\mathbf{P}^N\right)^2 = \mathbf{P}^N. \tag{5.3.33}$$

Second, it is a bounded linear operator in $L^2(\Omega)$ with operator norm unity. That is

$$\|\mathbf{P}^N\{f\}\|_2 \le \|f\|_2. \tag{5.3.34}$$

Third, it is self-adjoint. This means that for all square integrable f and g,

$$\int g^* \mathbf{P}^N \{f\}\, d^d x = \int \left(\mathbf{P}^N \{g\}\right)^* f\, d^d x. \tag{5.3.35}$$

Fourth, it commutes with derivatives for smooth functions:

$$\mathbf{P}^N \left\{ \frac{\partial f}{\partial x_j} \right\} = \frac{\partial}{\partial x_j} \mathbf{P}^N \{f\}. \tag{5.3.36}$$

The analogue of the energy evolution equation (5.3.18) is obtained in the continuum representation by dotting $\mathbf{u}^N$ into (5.3.27), integrating, and using the properties of the projector listed above. We find

$$\frac{1}{2}\frac{d}{dt}\|\mathbf{u}^N\|_2^2 = -\nu \|\nabla \mathbf{u}^N\|_2^2 + \int_\Omega \mathbf{u}^N \cdot \mathbf{f}\, d^d x, \tag{5.3.37}$$

which, up to an overall common factor, is exactly the same as (5.3.18). In this representation, the next step to the differential inequality in (5.3.19) involves applying Poincaré's inequality (Theorem 2.1) to the first term on the right-hand side,[1] and the Cauchy-Schwarz inequality to the second term on the right-hand side, yielding

$$\frac{1}{2}\frac{d}{dt}\|\mathbf{u}^N\|_2^2 \le -\nu \left(\frac{2\pi}{L}\right)^2 \|\mathbf{u}^N\|_2^2 + \|\mathbf{u}^N\|_2 \|\mathbf{f}\|_2. \tag{5.3.38}$$

Gronwall's lemma (Lemma 2.1) then implies that the total kinetic energy in the approximations, namely $\frac{1}{2}\|\mathbf{u}^N\|_2^2$, is bounded uniformly in time. It is also uniformly bounded in the order of the approximation N, because we can take the $N \to \infty$ limit on the right-hand side of (5.3.25) to obtain

$$\sum_{\mathbf{k}} \left|\hat{\mathbf{u}}^N(\mathbf{k},t)\right|^2 \le \max \left\{ \sum_{\text{all } \mathbf{k}} |\hat{\mathbf{u}}(\mathbf{k},0)|^2, \frac{L^2}{4\pi^2 \nu} \sum_{\text{all } \mathbf{k}} |\hat{\mathbf{f}}(\mathbf{k})|^2 \right\} < \infty. \tag{5.3.39}$$

This observation is the basis of the next step.

The so-called "weak" solutions of the Navier-Stokes equations are obtained as a limit of the Galerkin approximations as $N \to \infty$. The solutions are appropriately called "weak" for a number of reasons. For one, the sense of convergence of the approximations is in a weak topology. In the usual norm topology on a function space, a sequence of functions f_n converges to a limit f iff $\|f_n - f\| \to 0$ as $n \to \infty$. A weak topology on a function space is one where a sequence converges iff for every continuous linear scalar valued functional on the space, which we denote (f^*, f) with f^* a member of the dual space, the sequence obeys

[1] Recall that we are considering mean zero $\mathbf{u}^N$, so that Poincaré's inequality is valid.

$(f^*, f_n) \to (f^*, f)$. Convergence in the norm topology implies convergence in the weak topology, but not the reverse. The central feature of the weak topology used in establishing the existence of limiting weak solutions of the Navier-Stokes equations is the fact that it can be shown that the Galerkin approximations are in a compact set in an appropriate space of functions with an appropriate weak topology. A compact set enjoys the property that any sequence contained in it has a convergent subsequence. Hence there is a subsequence, $\mathbf{u}^{N_j}(\mathbf{x}, t)$, the limit of which, $\mathbf{u}(\mathbf{x}, t)$, we identify as a solution of the Navier-Stokes equations. At each instance of time the limit is a square integrable vector field. The existence of global weak solutions was first obtained by Leray over half a century ago, where the notion of weak solutions was first introduced. For the details of the functional settings and issues involved in establishing these limits, we refer the reader to the references.

The solutions so obtained are "weak" in another sense, too. In our treatment of ODEs in the previous section, the step taken after showing that Picard iterates converged was to show that the resulting limit satisfied the equation to which a solution was originally sought. There, this could be accomplished. For the weak solutions of the PDE, however, it can only be shown that the original PDE is satisfied in a "weak," or "smeared" sense. In particular, what follows from their existence proof is that for any (smooth) divergence-free vector field test function $\mathbf{v}(\mathbf{x})$, and times $t_2 > t_1 \geq 0$, the weak solutions $\mathbf{u}(\mathbf{x}, t)$ obey

$$\begin{aligned}\int_\Omega \mathbf{v}(\mathbf{x}) \cdot \mathbf{u}(\mathbf{x}, t_2)\, d^d x \;-\; \int_\Omega \mathbf{v}(\mathbf{x}) \cdot \mathbf{u}(\mathbf{x}, t_1)\, d^d x \\ = \int_{t_1}^{t_2} \int_\Omega [\nu \Delta \mathbf{v}(\mathbf{x})] \cdot \mathbf{u}(\mathbf{x}, t)\, d^d x\, dt \\ + \int_{t_1}^{t_2} \int_\Omega \mathbf{u}(\mathbf{x}, t) \cdot [\nabla \mathbf{v}(\mathbf{x})] \cdot \mathbf{u}(\mathbf{x}, t)\, d^d x\, dt. \qquad (5.3.40)\end{aligned}$$

This smeared relation follows easily from the Navier-Stokes equations if $\mathbf{u}(\mathbf{x}, t)$ is a solution in the usual sense – simply dot $\mathbf{v}(\mathbf{x})$ into the equations of motion and integrate over space and time, integrating by parts where necessary. This does *not* imply, however, that $\mathbf{u}(\mathbf{x}, t)$ actually satisfies the Navier-Stokes equations. By "the usual sense," used above, we mean that each term in the Navier-Stokes equations is well defined and smooth. However, just because $\mathbf{u}(\mathbf{x}, t)$ is, for instance, a square integrable function of space, it may not be the case that $\nu \Delta \mathbf{u}(\mathbf{x}, t)$ is square integrable as well. What does follow from the construction of weak solutions is that they are square integrable at each instant of time, and that the L^2 norms of their first derivatives are square integrable in bounded intervals of time. In fact, the weak solutions satisfy a form of the integrated energy equation known as *Leray's inequality*:

$$\frac{1}{2}\|\mathbf{u}(\cdot,t)\|_2^2 + \nu \int_0^t \|\nabla \mathbf{u}(\cdot,t')\|_2^2\,dt' \le \frac{1}{2}\|\mathbf{u}(\cdot,0)\|_2^2 + \int_0^t \int_\Omega \mathbf{u}(\mathbf{x},t')\cdot \mathbf{f}(\mathbf{x})\,d^dx\,dt'. \tag{5.3.41}$$

Note that it is the inequality in (5.3.41), rather than the equality that follows from a direct manipulation of the Navier-Stokes equations. This says that the kinetic energy at time t, plus the total energy dissipated up to time t, is not larger than the sum of the initial kinetic energy plus the total work performed by the body force, although it need not be equal to it. Hence the law of conservation of energy does not automatically follow from the construction of the weak solutions. We note, however, that there are no known examples of weak solutions which actually violate energy conservation.

Moreover, as discussed further in the next section, the weak solutions have not been shown to be unique in any sense. Their construction as the limit of a *subsequence* of the Galerkin approximations leaves open the possibility that there is more than one distinct limit, even for the same sequence of approximations. Nonunique evolution would violate the basic tenets of classical Newtonian determinism and would render the Navier-Stokes equations worthless as a predictive model.

What is missing for the weak solutions at this stage of the game is control over their smoothness. If the weak solutions were smooth enough that all of the terms in the Navier-Stokes equations made sense as normal functions, then we would say that the weak solutions were, in fact, "strong" solutions. To many readers this kind of distinction may seem little more than a mathematical formality of no real consequence or practical importance. The issues involved, however, go straight to the heart of the question of the validity and self-consistency of the Navier-Stokes equations as a hydrodynamic model, and the mathematical difficulties have their source in precisely the same physical phenomena that the equations are meant to describe.

5.4 Uniqueness and the regularity problem

In the first section of this chapter we developed an example ODE ($dz/dt = \sqrt{|z|}$) illustrating the kind of loss of unique forward evolution that is possible when a Lipschitz condition is missing. The same type of problem appears when we try to establish uniqueness of solutions for the Navier-Stokes equations. Suppose we formally try to show that solutions of the PDE are unique (forgetting for the moment that the weak solutions constructed in the last section don't necessarily literally satisfy the PDE). The way this is typically approached is to consider two solutions, $\mathbf{u}(\mathbf{x}, t)$ and $\mathbf{u}'(\mathbf{x}, t)$, both of which satisfy the equation and the same initial condition, and then to try to show that they must be equal. Toward this end, let us write down the equation that governs the difference between two solutions, $\mathbf{v} = \mathbf{u}' - \mathbf{u}$;

$$\frac{\partial \mathbf{v}}{\partial t} + \mathbf{v} \cdot \nabla \mathbf{v} + \mathbf{v} \cdot \nabla \mathbf{u} + \mathbf{u} \cdot \nabla \mathbf{v} + \nabla(p' - p) = \nu \Delta \mathbf{v}, \tag{5.4.1}$$

$$\nabla \cdot \mathbf{v} = 0. \tag{5.4.2}$$

The "energy" evolution equation for $\mathbf{v}$, obtained by dotting $\mathbf{v}$ into (5.4.1) and integrating, is

$$\frac{1}{2}\frac{d}{dt}\|\mathbf{v}\|_2^2 + \int \mathbf{v} \cdot (\nabla \mathbf{u}) \cdot \mathbf{v} d^d x = -\nu \|\nabla \mathbf{v}\|_2^2. \tag{5.4.3}$$

This can be recast as a differential inequality for $\|\mathbf{v}\|_2^2$ by applying Poincaré's inequality (Theorem 2.1) to the right-hand side and using the estimate

$$\left| \int \mathbf{v} \cdot (\nabla \mathbf{u}) \cdot \mathbf{v} \, d^d x \right| \leq \|\nabla \mathbf{u}\|_\infty \|\mathbf{v}\|_2^2, \tag{5.4.4}$$

where $\|\nabla \mathbf{u}(\cdot, t)\|_\infty$ is the maximum component of the supremum over space of $|\partial u_i, / \partial x_j|$ at the instant of time t. Then $\|\mathbf{v}\|_2^2$ obeys

$$\frac{1}{2}\frac{d}{dt}\|\mathbf{v}\|_2^2 \leq -\frac{4\pi^2 \nu}{L^2}\|\mathbf{v}\|_2^2 + \|\nabla \mathbf{u}(\cdot, t)\|_\infty \|\mathbf{v}\|_2^2. \tag{5.4.5}$$

Assuming that it makes sense, we can apply Gronwall's lemma (see Lemma 2.1) to deduce that

$$\|\mathbf{v}(\cdot, t)\|_2^2 \leq \|\mathbf{v}(\cdot, 0)\|_2^2 \exp\left\{ -\frac{4\pi^2 \nu t}{L^2} + \int_0^t \|\nabla \mathbf{u}(\cdot, t')\|_\infty \, dt' \right\}. \tag{5.4.6}$$

So *if* we knew that the argument of the exponent on the right-hand side of (5.4.6) was finite – in particular, if we knew that $\|\nabla \mathbf{u}(\cdot, t)\|_\infty$ had at most integrable singularities – then we could conclude that if $\mathbf{v}$ were initially equal to zero, then it would remain so. This would yield uniqueness. However, for the weak solutions we do not (yet) know that $\|\nabla \mathbf{u}(\cdot, t)\|_\infty$ remains so well behaved, and this illustrates the quandary that we are facing. The global weak solutions are square integrable at each instant of time, and their derivatives are square integrable in space and time, but these facts do not preclude singularities in the flow field across which $\|\nabla \mathbf{u}(\cdot, t)\|_\infty$ is not integrable in time. Roughly speaking as regards uniqueness, an integrability condition on $\|\nabla \mathbf{u}(\cdot, t)\|_\infty$ for the Navier-Stokes PDE is the analogue of a Lipschitz condition for an ODE. In order to infer uniqueness of solutions to the Navier-Stokes equations we must have more regularity than that provided directly by the process of constructing the weak solutions. For example, *if* we could show that each of the Galerkin approximations satisfied a bound of the form

$$\int_0^T \|\nabla \mathbf{u}^N(\cdot, t)\|_\infty \, dt \le C(T), \tag{5.4.7}$$

where $C(T)$ is a finite function of $T \ge 0$ which does *not* depend on N, then we could conclude that any limit of the $\mathbf{u}^N$, as $N \to \infty$, also would satisfy

$$\int_0^T \|\nabla \mathbf{u}(\cdot, t)\|_\infty dt \le C(T) < \infty. \tag{5.4.8}$$

Although control of the solutions of the form (5.4.7) or (5.4.8) does not obviously guarantee that higher derivatives of the solutions are necessarily well behaved, we will see in the next two chapters that this actually is the case. That is, if by some means we can prove that a bound of the form (5.4.7) or (5.4.8) holds then we will be able to produce explicit (finite) bounds on the norms of all derivatives of the velocity field. From this we may infer that the solutions are unique, smooth, infinitely differentiable functions. Solutions with these properties are called "classical" solutions of the Navier-Stokes equations.

The question of uniqueness of solutions is not the only motivation for inquiring into the regularity properties of solutions. If smoothness of solutions is lost on some level, then this implies the presence of small scale structures in the flow. A discontinuity implies a "macroscopic" change over an infinitesimal "microscopic" interval. On its own this may not lead to any difficulties, but the incompressible Navier-Stokes equations are derived from a microscopic model of interacting particles (atoms or molecules) precisely in a limit of infinite separation of length and time scales between the microscopic and the macroscopic phenomena (see the reference to Bardos, Golse, and Levermore in section 1.6). If the macroscopic hydrodynamic equations were then to generate singularities or structures at infinitesimal scales, this would signal a breakdown of the very *ansatz* leading to their derivation. Clearly this is a fundamental philosophical problem, but it is also a mathematical problem: Rigorous derivations of the hydrodynamic equations rely on the smoothness of their solutions to show convergence from the microscopic scales. So along with the question of uniqueness, it is important to monitor the length (and time) scales generated by the Navier-Stokes equations. As developed further in the next two chapters, the relevant technical issues are closely related.

The integrability condition on $\|\nabla \mathbf{u}(\cdot, t)\|_\infty$ in (5.4.8) is not the only sufficient condition for the solutions to be classical. If the derivatives are square integrable at each instant of time, i.e., if

$$\sup_{0 \le t \le T} \|\nabla \mathbf{u}^N(\cdot, t)\|_2^2 \le C(T) < \infty \tag{5.4.9}$$

and thus

$$\sup_{0 \le t \le T} \|\nabla \mathbf{u}(\cdot, t)\|_2^2 \le C(T) < \infty \tag{5.4.10}$$

then the same regularity properties may be inferred. (This is not intended to be obvious. Such a condition as (5.4.10) in itself does not preclude all possible singularities a priori, but when combined with the evolution equation, this inference actually can be established. This is accomplished in the following chapters.) The square of the L^2 norm of the derivatives of the velocity field is a physically relevant quantity to consider because, up to a proportionality factor (ν), it is the instantaneous rate of viscous energy dissipation. It is also a natural place to start to illustrate the essential differences between the $2d$ and $3d$ Navier-Stokes equations with respect to questions of uniqueness and regularity. Let us attempt to derive a bound like that in (5.4.9) for solutions of the Navier-Stokes equations on $[0, L]^d$ with periodic boundary conditions. We start with the PDE satisfied by the Galerkin approximation $\mathbf{u}^N$ as developed in section 5.3,

$$\frac{\partial \mathbf{u}^N}{\partial t} + \mathbf{P}^N \left\{ \mathbf{u}^N \cdot \nabla \mathbf{u}^N + \nabla p \right\} = \nu \Delta \mathbf{u}^N + \mathbf{P}^N \left\{ \mathbf{f}(\mathbf{x}) \right\}, \tag{5.4.11}$$

$$\nabla \cdot \mathbf{u}^N = 0, \tag{5.4.12}$$

and initial condition

$$\mathbf{u}^N(\mathbf{x}, 0) = \mathbf{P}^N \left\{ \mathbf{u}_0(\mathbf{x}) \right\}. \tag{5.4.13}$$

As a finite sum of terms of the form $e^{i\mathbf{k}\cdot\mathbf{x}}$, the unique global continuous solution $\mathbf{u}^N(\mathbf{x}, t)$ is an infinitely differentiable function, so we may freely manipulate it and its equation of motion with impunity, making no apologies and with no sacrifice of rigor.

Taking the gradient of (5.4.11) and (1) recalling that the self-adjoint operator $\mathbf{P}^N$ commutes with ∇, (2) contracting with $\nabla\mathbf{u}^N$, (3) integrating, and (4) integrating by parts, leads to the evolution equation for $\|\nabla\mathbf{u}^N\|_2^2$:

$$\frac{1}{2}\frac{d}{dt}\|\nabla\mathbf{u}^N\|_2^2 = -\nu\|\Delta\mathbf{u}^N\|_2^2 - \int u_{i,j}^N u_{i,k}^N u_{j,k}^N \, d^d x + \int \nabla\mathbf{u}^N \cdot \nabla\mathbf{f} \, d^d x. \quad (5.4.14)$$

The last term on the right-hand side may be estimated

$$\left|\int \nabla\mathbf{u}^N \cdot \nabla\mathbf{f} d^d x\right| \leq \|\nabla\mathbf{u}^N\|_2 \|\nabla\mathbf{f}\|_2, \quad (5.4.15)$$

and so in the following we will be assuming body force functions which (at least) have square integrable derivatives.[1]

For the middle term on the right-hand side of (5.4.14) first use the bound

$$\left|\int u_{i,j}^N u_{i,k}^N u_{j,k}^N d^d x\right| \leq \|\nabla\mathbf{u}\|_3^3 := \int \left(\sum_{i,j=1}^{d} u_{i,j}^2\right)^{3/2} d^d x \leq a_d \sum_{i,j=1}^{d} \int |u_{i,j}|^3 \, d^d x, \quad (5.4.16)$$

where a_d is a number which only depends on the spatial dimension d. Then we use one of the calculus inequalities of Gagliardo and Nirenberg (see the Appendix). In particular we use the fact that for all smooth functions of mean zero (valid for $d < 6$), there is a finite constant b_d so that

$$\|f\|_3 \leq b_d \|\nabla f\|_2^{d/6} \, \|f\|_2^{1-d/6}, \quad (5.4.17)$$

where

$$\|\nabla f\|_2^2 := \sum_{i=1}^{d} \left|\frac{\partial f}{\partial x_i}\right|^2 . \quad (5.4.18)$$

Using (5.4.16) and (5.4.17), we have

$$\left|\int u_{i,j}^N u_{i,k}^N u_{j,k}^N d^d x\right| \leq c_d \, \|\Delta\mathbf{u}^N\|_2^{d/2} \, \|\nabla\mathbf{u}^N\|_2^{(6-d)/2}, \quad (5.4.19)$$

where c_d is a finite number (for $d < 6$), and where we have used the fact[2] that the Laplacian in L^2 controls all second derivatives in L^2. That is,

[1] Although it is not always necessary, we will generally take body forces and initial conditions as smooth as we please. It would not be surprising if regularity depended on these quantities, but what we are interested in is the regularity – or lack of it – resulting from the action of the dynamics alone.

[2] This type of argument is discussed in greater detail in the next chapter in section 6.1.

there is finite number C_d so that for all smooth functions,

$$\sum_{i,j=1}^{d} \left| \frac{\partial^2 f}{\partial x_i \partial x_j} \right|^2 \leq C_d \, \|\Delta f\|_2^2. \tag{5.4.20}$$

(This fact is easy to prove using the Fourier transformed representation.)

Inserting the estimates in (5.4.15) and (5.4.19) into the evolution equation for $\|\nabla \mathbf{u}^N\|_2^2$ yields

$$\frac{1}{2}\frac{d}{dt}\|\nabla \mathbf{u}^N\|_2^2 \leq -\nu \|\Delta \mathbf{u}^N\|_2^2 + c_d \, \|\Delta \mathbf{u}^N\|_2^{d/2} \, \|\nabla \mathbf{u}^N\|_2^{(6-d)/2} + \|\nabla \mathbf{u}^N\|_2 \, \|\nabla \mathbf{f}\|_2. \tag{5.4.21}$$

In order to obtain a closed differential inequality for $\|\nabla \mathbf{u}^N\|_2^2$, the occurrences of $\|\Delta \mathbf{u}^N\|_2$ must be eliminated. For the second term on the right-hand side in spatial dimension $d < 4$, we use Hölder's inequality ($ab \leq a^p/p + b^q/q$, where $1/p + 1/q = 1$) with $p = 4/d$ and $q = 4/(4-d)$ to find

$$c_d \, \|\Delta \mathbf{u}^N\|_2^{d/2} \, \|\nabla \mathbf{u}^N\|^{(6-d)/2} \leq \nu \|\Delta \mathbf{u}^N\|_2^2 + c_d' \nu^{-d/(4-d)} \left(\|\nabla \mathbf{u}^N\|_2^2\right)^{\frac{6-d}{4-d}}, \tag{5.4.22}$$

where c_d' only depends on $d < 4$. Inserting (5.4.22) into (5.4.21) and canceling a factor of $\|\nabla \mathbf{u}^N\|_2$,

$$\frac{d}{dt}\|\nabla \mathbf{u}^N\|_2 \leq c_d' \nu^{-d/(4-d)} \|\nabla \mathbf{u}^N\|_2^{\frac{8-d}{4-d}} + \|\nabla \mathbf{f}\|_2. \tag{5.4.23}$$

Note that c_d' does not depend on the order of the Galerkin approximation, N, so that (5.4.23) is uniformly true for each of the approximations as long as $d < 4$, which includes the physically relevant cases of $d = 2$ and 3.

Now let us focus on the $d = 2$ case, where (5.4.23) becomes

$$\frac{d}{dt}\|\nabla \mathbf{u}^N\|_2 \leq c_2' \nu^{-1} \left(\|\nabla \mathbf{u}^N\|_2\right)^3 + \|\nabla \mathbf{f}\|_2 = c_2' \nu^{-1} \|\nabla \mathbf{u}^N\|_2^2 \times \|\nabla \mathbf{u}^N\|_2 + \|\nabla \mathbf{f}\|_2. \tag{5.4.24}$$

The Galerkin approximations satisfy the energy evolution equation (recall (5.3.37))

$$\frac{1}{2}\frac{d}{dt}\|\mathbf{u}^N\|_2^2 = -\nu \|\nabla \mathbf{u}^N\|_2^2 + \int_\Omega \mathbf{u}^N \cdot \mathbf{f} \, d^d x, \tag{5.4.25}$$

from which it follows that

$$\nu \int_0^t \|\nabla \mathbf{u}^N(\cdot, t')\|_2^2 \, dt' \leq \frac{1}{2}\|\mathbf{u}_0\|_2^2 + \frac{1}{2}\|\mathbf{f}\|_2^2 \, t, \tag{5.4.26}$$

where $\mathbf{u}_0(\mathbf{x}) = \mathbf{u}(\mathbf{x}, 0)$ is the initial condition. This bound on the time integral of $\|\nabla \mathbf{u}^N\|_2^2$ is uniform in N, and so survives the limiting procedure (it follows from Leray's inequality for the weak solutions). Gronwall's lemma (Lemma 2.1) applied to (5.4.24) yields

$$\|\nabla \mathbf{u}^N(\cdot,t)\|_2 \le \|\nabla \mathbf{u}^N(\cdot,0)\|_2 \exp\left\{\frac{c_2'}{\nu}\int_0^t \|\nabla \mathbf{u}^N(\cdot,t')\|_2^2\,dt'\right\} + \|\nabla \mathbf{f}\|_2 \int_0^t \exp\left\{\frac{c_2'}{\nu}\int_{t'}^t \|\nabla \mathbf{u}^N(\cdot,t'')\|_2^2\,dt''\right\} dt', \qquad (5.4.27)$$

and inserting the uniform bound in (5.4.26) and noting that $\|\nabla \mathbf{u}^N(\cdot,0)\|_2 \le \|\nabla \mathbf{u}_0\|_2$, which we presume is finite, we arrive at the explicit uniform (in N) finite upper bound on $\|\nabla \mathbf{u}^N(\cdot,t)\|_2$, valid for all $t \ge 0$,

$$\|\nabla \mathbf{u}^N(\cdot,t)\|_2 \le \|\nabla \mathbf{u}_0\|_2 \exp\left\{\frac{c_2'}{\nu^2}\left(\|\mathbf{u}_0\|_2^2 + \|\mathbf{f}\|_2^2\, t\right)\right\} + \|\nabla \mathbf{f}\|_2 \int_0^t \exp\left\{\frac{c_2'}{\nu^2}\left(\|\mathbf{u}_0\|_2^2 + \|\mathbf{f}\|_2^2 (t-t')\right)\right\} dt'. \qquad (5.4.28)$$

The conclusion: Any $N \to \infty$ limit of $\mathbf{u}^N$ must also have square integrable derivatives at each instant of time. Note that the viscosity is crucial. The upper bound in (5.4.28) degrades to no bound at all as ν is decreased to zero.

This is the first step of a full regularity (and thus also uniqueness) proof for the 2*d* Navier-Stokes equations on a periodic domain (i.e., in the absence of boundaries). In the next three chapters we will show that this result is sufficient to guarantee that all derivatives of any $N \to \infty$ limit of $\mathbf{u}^N$ are square integrable pointwise in time.

The situation is very different in three spatial dimensions. Returning to (5.4.23) and replacing d with 3, we have

$$\frac{d}{dt}\|\nabla \mathbf{u}^N\|_2 \le \frac{c_3'}{\nu^3}\left(\|\nabla \mathbf{u}^N\|_2\right)^5 + \|\nabla \mathbf{f}\|_2. \qquad (5.4.29)$$

If we were to attempt the same kind of application of Gronwall's inequality that worked in 2*d*, we would find it necessary to provide a uniform (in N) bound not just on the time integral of $\|\nabla \mathbf{u}^N\|_2^2$ – which is provided by Leray's inequality – but on the time integral of $\|\nabla \mathbf{u}^N\|_2^4$. To date, no such estimate exists. What we can do is to produce a bound $\|\nabla \mathbf{u}^N\|_2$ for short times. For simplicity let us set the forcing to zero. Consider the differential inequality

$$\frac{dh(t)}{dt} \le a\,[h(t)]^5 \qquad (5.4.30)$$

for a positive quantity h, where a is a positive constant. Then for positive infinitesimal dt, we have

$$\frac{dh}{h^5} \leq a\,dt. \tag{5.4.31}$$

Integrating, this gives

$$\int_{h(0)}^{h(t)} \frac{dh}{h^5} = \frac{1}{4}\left\{\frac{1}{[h(0)]^4} - \frac{1}{[h(t)]^4}\right\} \leq \int_0^t a\,dt' = at, \tag{5.4.32}$$

so as long as $t < \left[4ah(0)^4\right]^{-1}$,

$$h(t) \leq \frac{h(0)}{\left(1 - 4ah(0)^4 t\right)^{1/4}} < \infty. \tag{5.4.33}$$

Identifying $h(t)$ with $\|\nabla \mathbf{u}^N(\cdot,t)\|_2$ and a with c_3'/ν^3, we have (in the unforced case) a uniform (in N) bound for short times:

$$\begin{aligned} \|\nabla \mathbf{u}^N(\cdot,t)\|_2 &\leq \frac{\|\nabla \mathbf{u}^N(\cdot,0)\|_2}{\left(1 - 4c_3'\nu^{-3}\|\nabla \mathbf{u}^N(\cdot,0)\|_2^4\, t\right)^{1/4}} \\ &\leq \frac{\|\nabla \mathbf{u}_0\|_2}{\left(1 - 4c_3'\nu^{-3}\|\nabla \mathbf{u}_0\|_2^4\, t\right)^{1/4}}, \end{aligned} \tag{5.4.34}$$

which is valid for

$$t < T^* := \frac{\nu^3}{4c_3'\|\nabla \mathbf{u}_0\|_2^4}. \tag{5.4.35}$$

Thus if we start out with an initial condition with square integrable derivatives, then for times $t < T^*$ any limit of the Galerkin approximations will also have square integrable derivatives. After time T^*, we can say nothing. The divergence of the upper bound at T^* does not necessarily mean that $\|\nabla \mathbf{u}\|_2$ also diverges, only that we cannot be sure that it doesn't. Note that the length of time during which we are assured of a reasonable smooth solution, T^*, depends sensibly on the parameters of the problem: Increasing the viscosity lengthens T^*, as does smoothing the initial flow field (decreasing $\|\nabla \mathbf{u}_0\|_2$).

These results for the $2d$ and $3d$ Navier-Stokes equations summarize the current state of affairs. On a periodic domain, global weak solutions exist in both $2d$ and $3d$. A crucial distinction arises at the level $\|\nabla \mathbf{u}\|_2^2$, which is proportional to the instantaneous rate of the rate of energy dissipation and, for the periodic boundary conditions considered here, to the enstrophy $\|\omega\|_2^2$. In $2d$ these quantities are forever finite for solutions of the Navier-Stokes equations. In $3d$ they are bounded for short times, but even in the absence of boundaries and/or external forcing we do not know if they remain well behaved for arbitrarily long times.

The distinction between the 2*d* and 3*d* problems is made easier by considering the vorticity evolution equation for the Galerkin approximations. In the 2*d* case,

$$\frac{\partial \omega^N}{\partial t} + \mathbf{P}^N \left\{ \mathbf{u}^N \cdot \nabla \omega^N \right\} = \nu \Delta \omega^N + \mathbf{e}_3 \cdot \mathbf{P}^N \left\{ \operatorname{curl} \mathbf{f} \right\}, \tag{5.4.36}$$

where

$$\omega^N = \mathbf{e}_3 \cdot \left(\operatorname{curl} \mathbf{u}^N \right). \tag{5.4.37}$$

The enstrophy evolution equation is

$$\frac{1}{2}\frac{d}{dt}\|\omega^N\|_2^2 = -\nu \|\nabla \omega^N\|_2^2 + \int \omega^N \mathbf{e}_3 \cdot (\operatorname{curl} \mathbf{f}) d^2x, \tag{5.4.38}$$

and Poincaré's inequality (Theorem 2.1) applied to the first term on the right-hand side with the Cauchy-Schwarz inequality applied to the second term yields the differential inequality

$$\frac{d}{dt}\|\omega^N\|_2 = -\frac{4\pi^2\nu}{L^2}\|\omega^N\|_2 + \|\operatorname{curl} \mathbf{f}\|_2. \tag{5.4.39}$$

Gronwall's lemma (Lemma 2.1) then provides the uniform estimate (uniform in both N and t)

$$\begin{aligned} \|\omega^N(\cdot,t)\|_2 \le\ & \|\omega(\cdot,0)\|_2 \exp\left\{-\frac{4\pi^2\nu}{L^2}t\right\} \\ & + \frac{L^2}{4\pi^2\nu}\|\operatorname{curl} \mathbf{f}\|_2 \left(1 - \exp\left\{-\frac{4\pi^2\nu}{L^2}t\right\}\right). \end{aligned} \tag{5.4.40}$$

Hence the limit of the Galerkin approximations has finite enstrophy and a finite rate of energy dissipation pointwise in time.

The physical source of the 3*d* challenge is apparent in terms of the vorticity formulation. The 3*d* Galerkin approximations to the vorticity obey

$$\frac{\partial \omega^N}{\partial t} + \mathbf{P}^N \left\{ \mathbf{u}^N \cdot \nabla \omega^N \right\} = \nu \Delta \omega^N + \mathbf{P}^N \left\{ \omega^N \cdot \nabla \mathbf{u}^N \right\} + \mathbf{P}^N \left\{ \operatorname{curl} \mathbf{f} \right\}, \tag{5.4.41}$$

and the enstrophy evolution equation is

$$\frac{1}{2}\frac{d}{dt}\|\omega^N\|_2^2 = -\nu \|\nabla \omega^N\|_2^2 + \int \omega^N \cdot (\nabla \mathbf{u}^N) \cdot \omega^N d^3x + \int \omega^N \cdot (\operatorname{curl} \mathbf{f}) d^3x. \tag{5.4.42}$$

The problem comes precisely from the term associated with the 3*d* vortex stretching phenomena. If $\nabla \mathbf{u}^N$ was well behaved in space and time, uniformly in N, then there would be a chance of controlling the rate of enstrophy growth, specifically the second term on the right-hand side of

(5.4.42). Excessive values of $\nabla \mathbf{u}^N$ combined with the appropriate local orientation of $\boldsymbol{\omega}$ could, however, lead to singularly stretched vortices wherein the local angular velocity of a fluid element might diverge. The result could be a loss of smoothness of the velocity vector field, the appearance of infinitesimally small scale structures in the flow field, and perhaps even a loss of unique forward evolution. Whether or not these things take place in general for solutions to the $3d$ incompressible Navier-Stokes equations remains an open question.

5.5 References and further reading

The level of analysis presumed in this chapter does not exceed that in [20]. Function space topologies are discussed in [21] and [22]. Weak solutions of the Navier-Stokes equations were first constructed by Leray [23]. We emphasize that we have focused exclusively on the case of periodic boundary conditions in this chapter. Other situations are very different, for example (5.3.36) does not generally hold.

Modern treatments of Navier-Stokes analyses may be found in Constantin and Foias [12] and Temam [24].

Exercises

1 State and prove the contraction mapping theorem.

2 Prove, via induction from (5.2.7), that

$$\|\mathbf{z}_{n+1} - \mathbf{z}_n\| \le (KT)^n \, |\mathbf{F}(\mathbf{z}_0)| \, T. \tag{E5.1}$$

3 Derive (5.3.18) and (5.3.37) for the Galerkin approximations.

4 Prove that norm convergence implies weak convergence in L^2. Produce a counterexample to show that weak convergence does *not* imply norm convergence.

5 Prove (5.4.20) and estimate C_d.

6

Ladder results for the Navier-Stokes equations

6.1 Introduction

In the previous chapter it was shown how solutions of the Navier-Stokes equations could be constructed. Uniqueness of those solutions requires more regularity, however, than that which follows directly from their construction via the Galerkin approximations. In this chapter we shall begin to see how much regularity is needed to ensure smoothness of solutions of the Navier-Stokes equations. The minimum requirements can be reached for the $2d$ problem, but the problem remains open for the $3d$ case.

This chapter is devoted to the statement and proof of what will be referred to as the *ladder theorem* for the Navier-Stokes equations on $\Omega = [0, L]^d$ with periodic boundary conditions and zero mean, and a discussion of its consequences in both $2d$ and $3d$. This will enable us to relate the evolution of a seminorm of solutions of the Navier-Stokes equations, containing a given number of derivatives, to one containing a lower number of derivatives. In sections 6.3 and 6.4, it is shown how the ladder leads to the identification of length scales in the solutions. Subsequently, section 6.5 contains those estimates that can be gleaned via the ladder from the $2d$ and $3d$ Navier-Stokes equations where no assumptions have been made. Finally, to show how forcing fields can be handled differently from the static spatial forcing $\mathbf{f}(\mathbf{x})$ of previous chapters, section 6.6 briefly shows how a ladder may be derived for thermal convection.

To derive the ladder theorem, it is necessary to introduce the idea of seminorms which contain derivatives of the velocity field higher than unity. By now the L^2 norm of the velocity vector $\|\mathbf{u}\|_2^2$ and the sum of the L^2 norms of the gradient of each of the components of $\mathbf{u}$, namely

$\|\nabla\mathbf{u}\|_2^2$, should be familiar objects to the reader. It is now necessary to introduce the idea of the general derivative of the velocity field of order $N \geq 1$. In most books, the single derivative is called the D-operator.[1] Some books and papers still use the ∇ notation for D but this notation can be confusing. In terms of the D-operator, N derivatives on $\mathbf{u}$ in d dimensions is defined such that

$$|D^N\mathbf{u}|^2 = \sum_{|n|=N} \left|\frac{\partial^N \mathbf{u}}{\partial x_1^{n_1}\ldots\partial x_d^{n_d}}\right|^2, \tag{6.1.1}$$

where $|n| = n_1 + n_2 \ldots + n_d = N$. In other words $\left|D^N\mathbf{u}\right|^2$ is the sum of the squares of every derivative of order N. For example, in $2d$, using the subscript notation for partial derivatives, $\left|D^2\mathbf{u}\right|^2 = \sum_{i=1}^2 \left(u_{i,xx}^2 + 2u_{i,xy}^2 + u_{i,yy}^2\right)$. That is, we have the xx, xy, yx, and yy derivatives. Because the Laplacian operator Δ and the vector gradient ∇ explicitly appear in the Navier-Stokes equations and, additionally, because the curl operator appears in the vorticity $\boldsymbol{\omega}$, it is desirable to see how D, ∇ and curl are related. For instance, it is clear that whereas D^2 is a tensorlike quantity, the Laplacian Δ is a scalar operator. Clearly the two are not "equal." For this reason we shall spend a little time understanding the three operators and the relations between them.

The squares of seminorms H_N are defined by

$$H_N = \sum_{i=1}^{d} \int_\Omega \left|D^N u_i\right|^2 d^dx = \sum_{i=1}^{d} \int_\Omega \sum_{|n|=N} \left|\frac{\partial^N u_i}{\partial x_1^{n_1}\ldots\partial x_d^{n_d}}\right|^2 d^dx. \tag{6.1.2}$$

Recall that curl can be written in terms of the totally antisymmetric Levi-Cevita symbol ε_{ijk} as

$$\operatorname{curl}\mathbf{u} = \varepsilon_{ijk}\frac{\partial u_k}{\partial x_j}, \tag{6.1.3}$$

where

$$\varepsilon_{ijk}\varepsilon_{ilm} = \delta_{jl}\delta_{km} - \delta_{jm}\delta_{kl}. \tag{6.1.4}$$

To consider how seminorms of D, ∇, and curl are related on periodic

[1] In the book on PDEs by F. John, it is shown how the D-operator obeys all the usual properties enjoyed by a derivative such as integration by parts, a Leibnitz rule, etc. See the references section.

domains for divergence-free fields $\mathbf{u}$, we write

$$\begin{aligned}\int_\Omega |\mathrm{curl}\,\mathbf{u}|^2\, d^dx &= \int_\Omega \varepsilon_{ijk}\varepsilon_{ilm}u_{j,k}u_{l,m}\, d^dx \\ &= \int_\Omega \left(\delta_{jl}\delta_{km} - \delta_{jm}\delta_{kl}\right) u_{j,k}u_{l,m}\, d^dx \\ &= \int_\Omega \left[\left(\frac{\partial u_j}{\partial x_k}\right)^2 - \left(\frac{\partial u_j}{\partial x_k}\right)\left(\frac{\partial u_k}{\partial x_j}\right)\right] d^dx. \qquad (6.1.5)\end{aligned}$$

Integrating by parts twice on the last term produces a term $-u_{j,j}u_{k,k}$, which is zero because $\nabla\cdot\mathbf{u} = 0$. Hence

$$\int_\Omega |\mathrm{curl}\,\mathbf{u}|^2\, d^dx = \int_\Omega |\nabla\mathbf{u}|^2\, d^dx = \int_\Omega |D\mathbf{u}|^2\, d^dx = H_1. \qquad (6.1.6)$$

Hence the shorthand seminorm notation allows us to write

$$H_1 = \|\nabla\mathbf{u}\|_2^2 = \|D\mathbf{u}\|_2^2. \qquad (6.1.7)$$

Hence H_1 contains the sum of squares of the L^2 norms of the gradients of each component of the velocity field. Clearly, D and ∇ are equivalent, but how about D^2 and Δ? To establish the relationship between Δ, curl curl, and D^2, recall that $\mathrm{curl}\,\mathrm{curl}\,\mathbf{u} = -\Delta\mathbf{u} + \nabla\nabla\cdot\mathbf{u}$, which allows us to write

$$\int_\Omega |\mathrm{curl}\ \mathrm{curl}\,\mathbf{u}|^2\, d^dx = \int_\Omega |\Delta\mathbf{u}|^2\, d^dx. \qquad (6.1.8)$$

Now consider

$$\begin{aligned}\int_\Omega |\Delta\mathbf{u}|^2\, d^dx &= \int_\Omega \left|\sum_{j=1}^d \frac{\partial^2\mathbf{u}}{\partial x_j\partial x_j}\right|^2 d^dx \\ &= \int_\Omega \sum_{l,k=1}^d \left(\frac{\partial^2\mathbf{u}}{\partial x_l\partial x_l}\right)\cdot\left(\frac{\partial^2\mathbf{u}}{\partial x_k\partial x_k}\right) d^dx \\ &= \int_\Omega \sum_{l,k=1}^d \left|\frac{\partial^2\mathbf{u}}{\partial x_l\partial x_k}\right|^2 d^dx \\ &= \int_\Omega \left|D^2\mathbf{u}\right|^2 d^dx, \qquad (6.1.9)\end{aligned}$$

where we have integrated by parts twice to achieve the penultimate line. By generalizing these calculations, the reader can see that *while the operators D, curl and ∇ are not equivalent operators pointwise, nevertheless they are effectively equivalent under the L^2 norm.* The proof that

$$H_N = \int_\Omega \left|D^N\mathbf{u}\right|^2 d^dx = \int_\Omega \left|\mathrm{curl}^N\,\mathbf{u}\right|^2 d^dx \qquad (6.1.10)$$

for $N > 2$ is left as an exercise. However, this equivalence is not necessarily true[1] in every L^p except when $p = 2$. For instance, when $p = \infty$, let us look at $\|D\mathbf{u}\|_\infty$ and the L^∞ norm of the vorticity[2] $\|\omega\|_\infty$. While both these objects have the same dimension of frequency and may appear at first glance to be the same, in fact

$$\|\omega\|_\infty = \|\mathrm{curl}\,\mathbf{u}\|_\infty \neq \|D\mathbf{u}\|_\infty. \tag{6.1.11}$$

The reason is that $\|D\mathbf{u}\|_\infty$ is the supnorm of every derivative of every component of $\mathbf{u}$ while $\|\omega\|_\infty$ has derivatives of various components of $\mathbf{u}$ missing because of the way the curl operation is defined. It is eminently possible that the right-hand side of (6.1.11) could be infinite while the left-hand side is finite. We can say, however, that

$$\|\omega\|_\infty \leq \|D\mathbf{u}\|_\infty. \tag{6.1.12}$$

In the following sections we will appear to treat the Navier-Stokes equations and its solutions as though everything is defined classically. Really, the operations performed below should be considered to be applied to the Galerkin truncations and their associated equations of motion. None of the actual calculations depends on the order of the truncation, though, so we omit reference to it. The estimates produced are uniform in the order of the Galerkin approximation, and so hold for any limit of them, the weak solutions in particular.

6.2 The Navier-Stokes ladder theorem

In Chapter 1, the energy $2E(t) \equiv H_0 = \int_\Omega |\mathbf{u}|^2\, d^dx$ was shown to be bounded from above a priori in every dimension. Unfortunately, this tells us little about the velocity field $\mathbf{u}(\mathbf{x}, t)$ because functions bounded in L^2 can still display spatial singularities. To obtain more information about how the derivatives of the velocity and/or vorticity fields might be controlled, we need to consider the seminorms H_N defined in (6.1.2) and how they behave for arbitrary t. As equation (5.3.41) has shown, the only general information we have about the weak solutions is that the time integral of H_1 is bounded from above a priori (Leray's inequality). This does not tell us much about H_1 itself, as temporal singularities in the associated seminorm could still occur.

[1] Note also that certain terms have been discarded after the integrations by parts because of the periodic boundary conditions. It is also true that we have assumed that $\mathbf{u}$ is divergence free throughout.

[2] Although we use the notation $\|D\mathbf{u}\|_\infty$ in this, the next, and other chapters, it is interchangeable with $\|\nabla\mathbf{u}\|_\infty$.

It will turn out that control of the L^2 and the time average of the H_1 norms are crucial to our understanding of whether regularity can be achieved in both $2d$ and $3d$. It is desirable to have a theorem which relates H_N to H_{N-s} (say) for general N and $1 \leq s \leq N$. Toward this end, our aim in this chapter is to prove a theorem of this type. It is also necessary not only to show how the H_N evolve in time but also how the body forcing affects the flow. The theorem we will prove will be referred to as the "Navier-Stokes ladder theorem" or just "the ladder theorem." The main tools which will be used in the proof of this theorem are the Divergence Theorem, Cauchy's and Schwarz's inequalities, and finally the calculus inequalities of Gagliardo and Nirenberg listed in the Appendix. Let the set of seminorms associated with the forcing be defined by

$$\Phi_N = \int_\Omega \left|D^N \mathbf{f}\right|^2 d^d x, \tag{6.2.1}$$

where $\nabla \cdot \mathbf{f} = 0$ without loss of generality. Before we state this theorem it is necessary to make some restrictions on how the flow is forced. We assume a forcing which is independent of time, and in which the forcing function $\mathbf{f}(\mathbf{x})$ is assumed to have a cutoff in its spectrum such that it has a smallest length scale

$$\lambda_f^{-2} = sup_N \left(\frac{\Phi_{N+1}}{\Phi_N}\right). \tag{6.2.2}$$

In other words, the cutoff k_f in the spectrum is given by $k_f = 2\pi/\lambda_f$. To include the forcing, it is convenient to introduce a natural "time" $\tau = L^2\nu^{-1}$ for the system based on the length of the box L and the viscosity ν. The forcing function $\mathbf{f}$ is dimensionally an acceleration so $\mathbf{u_f} = \tau\mathbf{f}$ is dimensionally a velocity vector. Now we can define

$$F_N = H_N + \tau^2\Phi_N = H_N + \int_\Omega \left|D^N \mathbf{u_f}\right|^2 d^d x. \tag{6.2.3}$$

The F_N not only contains N derivatives on the three components of velocity but also the three components of the forcing. We define an inverse squared length scale as a combination of the inverse squared box length and inverse squared forcing length:

$$\lambda_0^{-2} = \lambda_f^{-2} + L^{-2}. \tag{6.2.4}$$

The following theorem is the main theorem of the chapter:

Theorem 6.1 *In $d = 2$ and $d = 3$ dimensions for periodic boundary conditions and for $1 \le s \le N$,*

$$\frac{1}{2}\dot{F}_N \le -\nu \frac{F_N^{1+1/s}}{F_{N-s}^{1/s}} + \left(c_N \|D\mathbf{u}\|_\infty + \nu\lambda_0^{-2}\right) F_N. \tag{6.2.5}$$

Moreover,

$$\frac{1}{2}\dot{F}_N \ge -\nu F_{N+1} - \left(c_N \|D\mathbf{u}\|_\infty + \nu\lambda_0^{-2}\right) F_N. \tag{6.2.6}$$

Proof: Before looking at the H_N, the following vector identity will prove useful

$$\nabla \cdot (\mathbf{A} \times \mathbf{B}) = \mathbf{B} \cdot \operatorname{curl} \mathbf{A} - \mathbf{A} \cdot \operatorname{curl} \mathbf{B}. \tag{6.2.7}$$

To begin, we differentiate the H_N with respect to time

$$\frac{1}{2}\dot{H}_N = \int_\Omega (\operatorname{curl}^N \mathbf{u}) \cdot \left[\operatorname{curl}^N (\nu\Delta\mathbf{u} - (\mathbf{u}\cdot\nabla)\mathbf{u} - \nabla p + \mathbf{f})\right] d^d x, \tag{6.2.8}$$

and then proceed with the proof of the theorem in nine steps.

STEP 1: The Laplacian term: Consider what happens to (6.2.7) under the effect of volume integration:

$$\int_\Omega \mathbf{B} \cdot \operatorname{curl} \mathbf{A}\, d^d x = \int_\Omega \mathbf{A} \cdot \operatorname{curl} \mathbf{B}\, d^d x. \tag{6.2.9}$$

The $\nabla \cdot (\mathbf{A} \times \mathbf{B})$ term vanishes under the application of the Divergence Theorem. Equation (6.2.9) is a type of integration by parts without the sign change. This rule enables us to deal with the Laplacian term in (6.2.8).

$$\text{Laplacian term} = \nu \int_\Omega \left(\operatorname{curl}^N \mathbf{u}\right) \cdot \left(\operatorname{curl}^N \Delta\mathbf{u}\right) d^d x. \tag{6.2.10}$$

Because $\operatorname{curl}\operatorname{curl}\mathbf{u} = -\Delta\mathbf{u}$, (6.2.9) can be used to move one of the curl operators across in (6.2.10). We therefore find

$$\text{Laplacian term} = -\nu \int_\Omega \left(\operatorname{curl}^N \mathbf{u}\right) \cdot \left(\operatorname{curl}^{N+2} \mathbf{u}\right) d^d x = -\nu H_{N+1}. \tag{6.2.11}$$

Note that (6.2.11) is an equality.

STEP 2: The nonlinear term: Here we use the D notation. This term becomes

$$|\text{Nonlinear term}| = \left| \int_\Omega \left(D^N \mathbf{u}\right) \cdot \left(D^N [-(\mathbf{u}\cdot\nabla)\mathbf{u}]\right) d^d x \right|. \tag{6.2.12}$$

Performing a Leibnitz expansion on the term with N derivatives and separating the first term in the expansion, the ith component is

$$|\mathrm{NLT}_i| \le -\int_\Omega \phi_i(\mathbf{u}\cdot\nabla)\phi_i\, d^dx + \left| \sum_{l=1}^{N}\sum_j C_l^N \int_\Omega \left(D^N u_i\right)\left(D^l u_j\right)\left(D^{N+1-l}u_i\right) d^dx \right|, \tag{6.2.13}$$

where $\phi_i = D^N u_i$. The first term on the RHS of (6.2.13) can be shown to be identically zero in the following way:

$$\int_\Omega \phi_i(\mathbf{u}\cdot\nabla)\phi_i\, d^dx = \int_\Omega \left[\nabla\cdot\left(\frac{1}{2}\phi_i^2\mathbf{u}\right) - \frac{1}{2}\phi_i^2\nabla\cdot\mathbf{u}\right] d^dx. \tag{6.2.14}$$

The last term is zero because $\nabla\cdot\mathbf{u} = 0$ and the first term is zero because of the Divergence Theorem and the periodic boundary conditions. After two applications of Hölder's inequality, with $1/p + 1/q = 1/2$, inequality (6.2.13) therefore becomes

$$|\text{NL term}| \le 2^N H_N^{1/2} \sum_{i,j}\sum_{l=1}^{N} \|D^l u_j\|_p\, \|D^{N+1-l}u_i\|_q. \tag{6.2.15}$$

The constant 2^N is the sum of the binomial coefficients. Since both $p, q > 2$, it is necessary to use calculus inequalities on the L^p and L^q norms in (6.2.15). Define $A_j \equiv Du_j$, and then use the calculus inequalities of the Appendix to obtain

$$\|D^{l-1}A_j\|_p \le c\, \|D^{N-1}A_j\|_2^a\, \|A_j\|_\infty^{1-a}. \tag{6.2.16}$$

In (6.2.16), the exponent a is calculated through the formula (A.0.20) which balances the dimensional weighting across the norms. From this

$$\frac{1}{p} = \frac{l-1}{d} + a\left(\frac{1}{2} - \frac{N-1}{d}\right), \tag{6.2.17}$$

where the value of a is restricted to be in the range

$$\frac{l-1}{N-1} \le a < 1. \tag{6.2.18}$$

The value of a is now deliberately chosen to be at the minimum of its range in inequality (6.2.18), in which case, for $l < N$, a is independent of the dimension d and takes the value

$$a = \frac{l-1}{N-1}, \qquad ap = 2. \tag{6.2.19}$$

The case $l = N$ does not require the use of (6.2.16). The same procedure on the L^q norm gives

$$\|D^{N-l}A_i\|_q \leq c\,\|D^{N-1}A_i\|_2^b\,\|A_i\|_\infty^{1-b}, \tag{6.2.20}$$

with ($l \geq 1$)

$$b = \frac{N-l}{N-1}, \qquad bq = 2. \tag{6.2.21}$$

Using the fact that $a + b = 1 \Rightarrow 1/p + 1/q = 1/2$, the combination of these results in (6.2.15) means that we have proved the following lemma which will be useful later on;

Lemma 6.1 *An estimate for the integral*

$$I_{ij} = \left|\int_\Omega (D^N u_i)(D^l u_j)(D^{N+1-l}u_i)\,d^d x\right| \tag{6.2.22}$$

is given by

$$I_{ij} \leq c_N \left(H_N^{(i)}\right)^{1/2+1/q}\left(H_N^{(j)}\right)^{1/p}\|Du_j\|_\infty^{1-a}\,\|Du_i\|_\infty^{1-b}, \tag{6.2.23}$$

where

$$H_N^{(i)} = \int_\Omega \left|D^N u_i\right|^2 d^d x. \tag{6.2.24}$$

Continuing with Step 2 of the theorem proof, we note that because $H_N^{(i)} \leq H_N$, $\|Du_i\|_\infty \leq \|\mathbf{Du}\|_\infty$ and $a + b = 1$, we end up with a simple upper bound for inequality (6.2.15)

$$|\text{NL term}| \leq c_N\,H_N\|\mathbf{Du}\|_\infty. \tag{6.2.25}$$

Notice that the constants c_N are dimensionless and that the RHS of (6.2.25) has the same dimension as that of $\dot{H}_N$. This is the end of Step 2.

STEP 3: The pressure term: The pressure term vanishes identically by the action of the curl operation in H_N.

STEP 4: The forcing term: This is the last term in (6.2.8) and, after an application of the Cauchy-Schwarz inequality, becomes

$$\left|\int_\Omega \left(\text{curl}^N\mathbf{u}\right)\cdot\left(\text{curl}^N\mathbf{f}\right)\,d^d x\right| \leq H_N^{1/2}\Phi_N^{1/2}. \tag{6.2.26}$$

The Schwarz inequality has been used.

STEP 5: The combination of Steps 1–4: Let us now put all these results together

$$\frac{1}{2}\dot{H}_N \leq -\nu H_{N+1} + c_N H_N \|D\mathbf{u}\|_\infty + H_N^{1/2}\Phi_N^{1/2}. \qquad (6.2.27)$$

STEP 6: An upper bound on $-H_{N+1}$: This can be achieved through the following

Lemma 6.2 *For $1 \leq s \leq N$ and $r \geq 1$, the H_N defined in (6.1.2) satisfy the inequality*

$$H_N \leq H_{N-s}^{\frac{r}{r+s}} H_{N+r}^{\frac{s}{r+s}}. \qquad (6.2.28)$$

Proof: This proceeds by induction and is performed in four steps A to D.

Step A: Let us take the definition of M_N as

$$M_N = \int_\Omega |D^N \mathbf{v}|^2 \, d^d x, \qquad (6.2.29)$$

and, after an integration by parts and the Cauchy-Schwarz inequality, obtain

$$M_N \leq \left(\int_\Omega \left|D^{N-1}\mathbf{v}\right|^2 d^d x\right)^{1/2} \left(\int_\Omega \left|D^{N+1}\mathbf{v}\right|^2 d^d x\right)^{1/2}. \qquad (6.2.30)$$

Hence we obtain

$$M_N \leq M_{N+1}^{1/2} M_{N-1}^{1/2}. \qquad (6.2.31)$$

This is also true separately for each component.

Step B: Second, we show that $\forall\, s \geq 1$,

$$M_s \;\leq\; M_{s+1}^{\frac{s}{(s+1)}} \, M_0^{\frac{1}{(s+1)}}. \qquad (6.2.32)$$

We know from (6.2.31) that (6.2.32) holds for $s = 1$. Assume (6.2.32) holds for $s > 1$. Then

$$M_{s+1} \;\leq\; M_{s+2}^{1/2} M_s^{1/2} \;\leq\; M_{s+2}^{1/2} M_{s+1}^{\frac{s}{2(s+1)}} M_0^{\frac{1}{2(s+1)}}, \qquad (6.2.33)$$

so

$$M_{s+1} \;\leq\; M_{s+2}^{\frac{(s+1)}{(s+2)}} M_0^{\frac{1}{(s+2)}}. \qquad (6.2.34)$$

Hence (6.2.32) is true $\forall\, s$ by induction.

Step C: Third, we show that $\forall\, s, r$

$$M_s \;\le\; M_{s+r}^{\frac{s}{(s+r)}}\, M_0^{\frac{r}{(s+r)}}. \tag{6.2.35}$$

We know from (6.2.32) that (6.2.35) holds for $r = 1$. Assume (6.2.35) holds for $r > 1$. Then

$$M_s \;\le\; M_{s+r}^{\frac{s}{(s+r)}}\, M_0^{\frac{r}{(s+r)}} \;\le\; M_{s+r+1}^{\frac{s}{(s+r+1)}}\, M_0^{\frac{(r+1)}{(s+r+1)}}, \tag{6.2.36}$$

where we have used (6.2.32). Hence (6.2.35) is true $\forall\, s, r$, by induction.

Step D: We know (6.2.35) holds with

$$M_s \;=\; \|D^s \mathbf{v}\|_2^2. \tag{6.2.37}$$

Now suppose that $v_i = D^{N-s}u_i$, then $M_s = H_N$ and $M_0 = H_{N-s}$. Equation (6.2.35) becomes

$$H_N \;\le\; H_{N+r}^{\frac{s}{(s+r)}}\, H_{N-s}^{\frac{r}{(s+r)}}. \tag{6.2.38}$$

Clearly, the choice $r = 1$ in the above lemma gives the required term in the theorem.

We have now demonstrated that for $1 \le s \le N$ and in dimension d, the H_N satisfy

$$\frac{1}{2}\dot{H}_N \le -\nu \frac{H_N^{1+1/s}}{H_{N-s}^{1/s}} + c_N\, \|D\mathbf{u}\|_\infty H_N + H_N^{1/2}\Phi_N^{1/2}. \tag{6.2.39}$$

This is a form of ladder theorem for the H_N. It is inconvenient to leave it in this form because of the way the forcing term is expressed here. Later on, we will want to divide through by H_N and the square root in the forcing term will cause complications. For this reason we set about dealing with the forcing term.

STEP 7: Returning to the theorem proof at the stage of (6.2.27), we can add and subtract a $\nu\tau^2\Phi_{N+1}$ term to obtain

$$\frac{1}{2}\dot{F}_N \le -\nu F_{N+1} + c_N\, F_N \|D\mathbf{u}\|_\infty + \nu\tau^2\Phi_{N+1} + F_N^{1/2}\Phi_N^{1/2}, \tag{6.2.40}$$

which can be rewritten as

$$\frac{1}{2}\dot{F}_N \le -\nu F_{N+1} + \left(c_N \|D\mathbf{u}\|_\infty + \nu F_N^{-1}\tau^2\Phi_{N+1} + \tau^{-1}\right) F_N. \tag{6.2.41}$$

Since from (6.2.3)

$$F_N \ge \tau^2\Phi_N, \tag{6.2.42}$$

we can appeal to (6.2.2), (6.2.4), and (6.2.41) can be written as

$$\frac{1}{2}\dot{F}_N \le -\nu F_{N+1} + \left(c_N \|D\mathbf{u}\|_\infty + \nu\lambda_0^{-2}\right) F_N, \tag{6.2.43}$$

with λ_0 defined as in (6.2.4). We now proceed to Step 8.

STEP 8: The result of Lemma 6.2 which we established for $s \ge 1$, $r = 1$ is also true for every "component" of $H_N = \sum_{i=1} H_N^{(i)}$ where the suffix i refers to the ith component u_i of the velocity field:

$$\left(H_N^{(i)}\right)^{1+s} \le \left(H_{N+1}^{(i)}\right)\left(H_{N-s}^{(i)}\right)^s. \tag{6.2.44}$$

We can therefore take the $(s+1)$th root and sum over i from 1 to 6 which, in effect, includes all 6 "components" of F_N: three from H_N and three from the forcing. Therefore

$$F_N = \sum_{i=1}^{6}\left(H_{N+1}^{(i)}\right)^{\frac{s}{1+s}}\left(H_{N-s}^{(i)}\right)^{\frac{1}{1+s}}. \tag{6.2.45}$$

Using the Hölder-Schwarz inequality, we have proved

Lemma 6.3 *For $1 \le s \le N$, the F_N defined in (6.2.3) satisfy the inequality*

$$F_N \le F_{N-s}^{\frac{1}{1+s}} F_{N+1}^{\frac{s}{1+s}}. \tag{6.2.46}$$

This gives the first term on the right hand side of (6.2.5).

STEP 9: Proof of (6.2.6): Step 1, which concerns the Laplacian term, contains no inequalities: It is an equality. In consequence, we can bound the time derivative of H_N below as well as above

$$\frac{1}{2}\dot{H}_N \ge -\nu H_{N+1} - c_N H_N \|D\mathbf{u}\|_\infty - H_N^{1/2}\Phi_N^{1/2}. \tag{6.2.47}$$

Finally, to obtain the last step in the proof of (6.2.6), and noting that $H_{N+1} \le F_{N+1}$, it is not difficult to see that

$$\frac{1}{2}\dot{F}_N \ge -\nu F_{N+1} - c_N \|D\mathbf{u}\|_\infty F_N - \tau^{-1} F_N, \tag{6.2.48}$$

where $\tau = L^2\nu^{-1}$. Because $\tau^{-1} \le \nu\lambda_0^{-2}$, the result follows. □

6.3 A natural definition of a length scale

Computing upper bounds on norms of the Navier-Stokes velocity field $\mathbf{u}(\mathbf{x}, t)$ is a vacuous exercise if pursued without recourse to physical interpretation. The ultimate aim of these calculations is to produce a lower bound on the smallest length scale in a forced flow. The term "smallest length scale" is not uniquely defined and so requires some thought. Ideally, we would like to resolve the smallest "eddy" in the flow starting from smooth initial conditions. To see how Theorem 6.1 naturally defines a set of length scales, we show that the F_N ladder suggests length scales associated with moments of the power spectrum.

A scale in turbulence theory may be defined in the following way. From the energy in the flow

$$2E(t) \equiv H_0 = \int_\Omega |\mathbf{u}(\mathbf{x}, t)|^2 d^d x, \tag{6.3.1}$$

we can use Parseval's theorem

$$E(t) = \frac{1}{2} \int |\hat{\mathbf{u}}(\mathbf{k}, t)|^2 d^d k \equiv \int E(\mathbf{k}, t) d^d k \tag{6.3.2}$$

to define the instantaneous energy spectrum $E(\mathbf{k}, t)$. This defines the distribution of energy among scales and also allows us to consider the normalized "probability distribution"

$$P(\mathbf{k}, t) = \frac{E(\mathbf{k}, t)}{E(t)}. \tag{6.3.3}$$

Time-dependent moments of this distribution can then be written as

$$\left[k^{2N}\right]_{s.a.} = \frac{\int |k|^{2N} |\hat{\mathbf{u}}(\mathbf{k}, t)|^2 d^d k}{\int |\hat{\mathbf{u}}(\mathbf{k}, t)|^2 d^d k} := \frac{H_N}{H_0}, \tag{6.3.4}$$

where the *s.a.* on $\left[k^{2N}\right]_{s.a.}$ stands for "spatial average." The relevant time-dependent length scale associated with this quantity is

$$(\text{length})^{-1} \sim \left\{ \left[k^{2N}\right]_{s.a.} \right\}^{1/2N}. \tag{6.3.5}$$

Not only can the forcing be included in the definitions to make the H_N into F_N, but the ratio of moments

$$\frac{\left[k^{2N}\right]_{s.a.}}{\left[k^{2r}\right]_{s.a.}} = \frac{F_N}{F_r} \tag{6.3.6}$$

also naturally defines a length scale

$$(\text{length})^{-1} \sim \left(\frac{F_N}{F_r} \right)^{\frac{1}{2(N-r)}}. \tag{6.3.7}$$

Clearly, we can define a set of "wavenumbers" defined by the ratio of moments above.

Definition 6.1

$$\kappa_{N,r}(t) = \left(\frac{F_N}{F_r}\right)^{\frac{1}{2(N-r)}}, \qquad r = N - s \geq 0. \tag{6.3.8}$$

It is noteworthy that if one divides the ladder inequality in Theorem 6.1 through by F_N, then the square of our time dependent "wavenumbers" $\kappa_{N,r}$ appear in the Laplacian term. It is convenient and natural, therefore, to time average the square of these objects to get an associated set of inverse squared lengths.

Definition 6.2

$$\ell_{N,r}^{-2} \equiv \langle \kappa_{N,r}^2 \rangle = \left\langle \left(\frac{F_N}{F_r}\right)^{\frac{1}{(N-r)}} \right\rangle, \tag{6.3.9}$$

where the time average $\langle g(t) \rangle$ is

$$\langle g(t) \rangle = \sup_{g(0)} \limsup_{t \to \infty} \frac{1}{t} \int_0^t g(\tau) d\tau. \tag{6.3.10}$$

For the moment, it is instructive to see how our definition of a length scale compares with the way time averages of moments are taken to define a length scale in turbulence theory. There, it is usual to consider the time averaged energy spectrum $\langle E(\mathbf{k}) \rangle$. The relevant length scales are defined in terms of the average distribution of energy

$$\langle P \rangle (\mathbf{k}) = \frac{\langle E(\mathbf{k}, \cdot) \rangle}{\int d^d k' \langle E(\mathbf{k}', \cdot) \rangle}. \tag{6.3.11}$$

This, of course, is not the same as time averaged instantaneous distribution of energy

$$\langle P(\mathbf{k}, \cdot) \rangle = \left\langle \frac{E(\mathbf{k}, \cdot)}{\int d^d k' E(\mathbf{k}', \cdot)} \right\rangle. \tag{6.3.12}$$

A conventional definition of a length scale is normally calculated via $\langle P \rangle (\mathbf{k})$ and so a *ratio of time averages* of the F_N and F_r would occur. For the scale given in Definition 6.2 we have a *time average of a ratio*. Moments computed by taking the time average last are going to be more sensitive to rare fluctuations driving energy deep down to shorter scales than those computed from the time averaged energy spectrum. This is because, all other things being equal, rare events contribute little to the time averaged energy spectrum. Moreover, when a significant fraction of energy is at high wavenumbers, the (relative) energy dissipation is necessarily higher, so these events will typically be characterized by lower than normal total energy. They will then not even count that much toward the time average of the energy spectrum. Dividing by the total energy *before* averaging therefore amplifies the role of the low energy but high wavenumber configurations in the distribution, skewing the distribution toward high wavenumbers.[1] Consequently, we therefore assert that the natural scale defined by computing averages on the $\kappa_{N,r}^2$ are likely to be more sensitive to intermittent fluctuations than a length scale determined by the time averaged energy spectrum alone. Because of the operation of the time average, however, it is still possible that we may not have resolved all length scales in the fluid, a problem to which we will return in later chapters.

6.4 The dynamical wavenumbers $\kappa_{N,r}$

Having talked above of computing the time average of ratios it is now time to see how such quantities can be obtained from the ladder. Dividing through equation (6.2.5) in the ladder theorem by F_N and time averaging according to Definition (6.2), we find

$$\frac{1}{2\nu}\left\langle \frac{d}{dt}\left[\log F_N\right]\right\rangle + \left\langle \kappa_{N,r}^2\right\rangle \leq c_N\, \nu^{-1}\left\langle \|\mathbf{Du}\|_\infty\right\rangle + \lambda_0^{-2}. \tag{6.4.1}$$

The integral in the average $\langle\cdot\rangle$ integrates the time derivative perfectly leaving the left-hand side as $\log F_N(t) - \log F_N(0)$. For each N, provided

$$\int_\Omega \left|D^N \mathbf{u_f}\right|^2 d^dx > 1, \tag{6.4.2}$$

F_N is bounded below by a fixed positive constant dependent only on the forcing and so the $\log F_N$ term is always bounded from below. These terms are therefore bounded by something that vanishes in the large time limit because of the $1/t$ in Definition 6.2 of the time average. We have proved

Theorem 6.2 *For each value of N, if $\|D^N\mathbf{u_f}\|_2^2 > 1$, then*

$$\left\langle \kappa_{N,r}^2\right\rangle \leq c_N\, \nu^{-1}\left\langle \|\mathbf{Du}\|_\infty\right\rangle + \lambda_0^{-2}. \tag{6.4.3}$$

□

[1] Nevertheless, see section 7.7 for estimates for a ratio of time averages.

Remark: Note that in its most important part, the $\langle \|D\mathbf{u}\|_\infty \rangle$ term, the RHS of (6.4.3) is ostensibly independent of the domain length L.

Because we have a ladder theorem for the time evolution of F_N and F_r, it is worth seeing how the $\kappa_{N,r}$ also evolve with respect to time:

$$2(N-r)\dot{\kappa}_{N,r}\kappa_{N,r}^{2(N-r)-1} = \frac{\dot{F}_N}{F_r} - \frac{F_N}{F_r}\frac{\dot{F}_r}{F_r}. \tag{6.4.4}$$

We obviously need the result of Theorem 6.1 for an upper bound on $\dot{F}_N$ and (6.2.6) for a lower bound on $\dot{F}_r$. In terms of $\kappa_{N,r}$, we obtain

$$\frac{1}{2}(N-r)\,\dot{\kappa}_{N,r} \leq -\frac{\nu}{2}\kappa_{N,r}\left(\kappa_{N,r}^2 - \kappa_{r+1,r}^2\right) + \left(c_N\,\|D\mathbf{u}\|_\infty + \nu\lambda_0^{-2}\right)\kappa_{N,r}. \tag{6.4.5}$$

The consequences of (6.4.5) in the $2d$ and $3d$ cases are discussed in Chapter 7. The inequality in (6.4.5) also gives us a clue about how to get some results on the nature of $3d$ Euler singularities, a topic which we will also address in Chapter 7.

Turning back to Definition 6.1 for $\kappa_{N,r}$ and noting that it has a dimension of inverse length, we remark that it is also possible to show that it has some meaning in terms of the calculus inequalities of the Appendix, which have already been used in the proof of the ladder theorem. If we take the rth derivative of u, namely $D^r u$, and ask what the calculus inequality looks like which interpolates between $\|D^r u\|_\infty$ and $\|D^r u\|_2$, we ought to obtain a quantity which has dimension (inverse length)$^{d/2}$ because of compensation for the volume integral in the L^2 norm.

$$\|D^r u_i\|_\infty \leq c\,\|D^N u_i\|_2^a\,\|D^r u_i\|_2^{(1-a)}, \tag{6.4.6}$$

with $a = \frac{d}{2(N-r)}$ and $N > r + d/2$. To form $\kappa_{N,r}$, we notice that $\|D^N u_i\|_2^2 \leq F_N$ and $\|D^r u_i\|_2^2 \leq F_r$. Dividing through (6.4.6) by $F_r^{a/2}$, we find that

$$\|D^r u_i\|_\infty \leq c\,\kappa_{N,r}^{d/2}\,\|D^r u_i\|_2. \tag{6.4.7}$$

Here we see that $\kappa_{N,r}^{d/2}$ is the object that has the dimension of (inverse length)$^{d/2}$. *Another way of looking at it is to say that $\kappa_{N,r}$ mediates between the L^∞ and L^2 norms of the quantity $D^r u_i$.* It is not too hard (see Exercises) to show that the $\kappa_{N,r}$ are ordered in increasing values of N for fixed r and increasing values of r for fixed N.

6.5 Estimates for the Navier-Stokes equations

Summarizing, the results of the previous sections may be tabulated as shown in Table 6.1

Table 6.1. *A summary of results so far. See the exercises for a reference to the second F_N ladder*

Definition of F_N	$F_N = \left(\lVert D^N \mathbf{u}\rVert_2^2 + \lVert D^N \mathbf{u}_\mathbf{f}\rVert_2^2\right)$
1st F_N ladder	$\frac{1}{2}\dot{F}_N \leq -\nu \frac{F_N^{1+1/s}}{F_{N-s}^{1/s}} + \left(c\,\lVert D\mathbf{u}\rVert_\infty + \nu\lambda_0^{-2}\right) F_N$
2nd F_N ladder	$\frac{1}{2}\dot{F}_N \leq -\frac{\nu}{2} \frac{F_N^{1+1/s}}{F_{N-s}^{1/s}} + \left(c\,\nu^{-1}\lVert \mathbf{u}\rVert_\infty^2 + \nu\lambda_0^{-2}\right) F_N$
Definition of $\kappa_{N,r}$	$\kappa_{N,r} = \left(\frac{F_N}{F_r}\right)^{\frac{1}{2(N-r)}} \qquad r < N$
Time average of $\kappa_{N,r}^2$	$\langle \kappa_{N,r}^2 \rangle \leq c_{N,r}\,\nu^{-1}\langle \lVert D\mathbf{u}\rVert_\infty \rangle + \lambda_0^{-2}$
Ladder for $\kappa_{N,r}$	$\frac{(N-r)}{2}\dot{\kappa}_{N,r} \leq -\frac{\nu}{2}\left(\kappa_{N,r}^2 - \kappa_{r+1,r}^2\right)\kappa_{N,r} + \left(c\,\lVert D\mathbf{u}\rVert_\infty + \nu\lambda_0^{-2}\right)\kappa_{N,r}$

Although the two versions of the ladder given in Table 6.1 above are the most general results for arbitrary N, for the three lowest rungs ($N = 0, 1$, and 2), there are more sensitive estimates for F_0, F_1, and F_2. Having good bounds on these is essential as the ladder is expressed as a type of a recurrence relation. In previous chapters, it has already been seen how the energy, $\int_\Omega |\mathbf{u}|^2\, d^d x$, and the enstrophy, $\int_\Omega |\boldsymbol{\omega}|^2\, d^d x$ introduced in section 3.2, are quantities which are important in both $2d$ and $3d$ flows, although for very different reasons.

6.5.1 Estimates for F_0

For a divergence-free flow, let us again write down the Navier-Stokes equations

$$\mathbf{u}_t + \mathbf{u}\cdot\nabla\mathbf{u} = \nu\Delta\mathbf{u} - \nabla p + \mathbf{f}, \tag{6.5.1}$$

with $\nabla\cdot\mathbf{u} = 0$. The only level where it is not necessary to use our standard F_N-notation is at $N = 0$. Using the vector identity given in (E1.1), the Navier-Stokes equations in (6.5.1) can be rewritten as

$$\mathbf{u}_t + \boldsymbol{\omega}\times\mathbf{u} = \nu\Delta\mathbf{u} - \nabla\left(p + \frac{1}{2}u^2\right) + \mathbf{f}. \tag{6.5.2}$$

Taking the scalar product of $\mathbf{u}$ with (6.5.2) and integrating over the domain gives the evolution of the energy $\frac{1}{2}\lVert\mathbf{u}\rVert_2^2$ (as in (1.4.21)). The nonlinear term immediately vanishes and the energy evolves according to

$$\frac{d}{dt}\left(\frac{1}{2}\lVert\mathbf{u}\rVert_2^2\right) = -\nu\lVert\nabla\mathbf{u}\rVert_2^2 + \int_\Omega \mathbf{u}\cdot\mathbf{f}\, d^d x. \tag{6.5.3}$$

An appeal to Poincaré's inequality in Theorem 2.1, which is used on the Laplacian term, and use of the Cauchy-Schwarz inequality on the forcing term, give

$$\frac{d}{dt}\left(\frac{1}{2}\|\mathbf{u}\|_2^2\right) \leq -\nu k_1^2\|\mathbf{u}\|_2^2 + \|\mathbf{u}\|_2\|\mathbf{f}\|_2, \tag{6.5.4}$$

where $k_1 = 2\pi/L$. To express the $\|\mathbf{u}\|_2^2$ in terms of the forcing is easier if we use the dimensionless Grashof numbers defined in (2.1.9):

$$2d \qquad \mathscr{G} = \frac{L^2\|\mathbf{f}\|_2}{\nu^2}, \qquad\qquad 3d \qquad \mathscr{G} = \frac{L^{3/2}\|\mathbf{f}\|_2}{\nu^2}. \tag{6.5.5}$$

Then, from (6.5.4), limsup estimates for $\|\mathbf{u}\|_2^2$ are given by (using Gronwall's lemma 2.1)

$$2d \qquad \overline{\lim}_{t\to\infty}\|\mathbf{u}\|_2^2 \leq c\,\nu^2\mathscr{G}^2, \qquad 3d \qquad \overline{\lim}_{t\to\infty}\|\mathbf{u}\|_2^2 \leq c\,L\nu^2\mathscr{G}^2, \tag{6.5.6}$$

where the dimensionless constants are denoted by c. The energy estimates given in (6.5.6) above, while useful, are only part of what we need in order to find an estimate for F_0. Because all the F_N contain forcing terms, estimates for these must also be included. From (6.2.3), we recall that

$$F_0 = \|\mathbf{u}\|_2^2 + \nu^{-2}L^4\|\mathbf{f}\|_2^2. \tag{6.5.7}$$

It is desirable to express the extra term $\nu^{-2}L^4\|\mathbf{f}\|_2^2$ in terms of the Grashof numbers defined above. From (6.2.1) we recall that our choice of forcing function is such that it possesses a smallest scale (see (6.2.2)), called λ_f, which appears in the definition of the standard length scale λ_0 (see (6.2.4))

$$\lambda_0^{-2} = L^{-2} + \lambda_f^{-2}. \tag{6.5.8}$$

When the forcing is included, (6.5.6) becomes

$$2d \qquad \overline{\lim}_{t\to\infty}F_0 \leq c\,\nu^2\mathscr{G}^2, \qquad 3d \qquad \overline{\lim}_{t\to\infty}F_0 \leq c\,L\nu^2\mathscr{G}^2. \tag{6.5.9}$$

These results are listed in Tables 6.2 and 6.3, (section 6.5.3), the first of which is for $2d$ results and the second for $3d$ results.

6.5.2 Estimates for $\langle F_1\rangle$ and $\langle\kappa_{1,0}^2\rangle$

The ladder for $N = 0$ is

$$\frac{1}{2}\dot{F}_0 \leq -\nu F_1 + \nu\lambda_0^{-2}F_0. \tag{6.5.10}$$

Time averaging (6.5.10) and using (6.5.6) therefore produces

$$2d \qquad \langle F_1\rangle \leq c\,\lambda_0^{-2}\nu^2\mathscr{G}^2, \qquad 3d \qquad \langle F_1\rangle \leq c\,L\lambda_0^{-2}\nu^2\mathscr{G}^2. \tag{6.5.11}$$

The $3d$ estimate above, in particular, is the time averaged version of Leray's inequality (see (5.3.41)). Note that our estimate for F_1 in (6.5.11) includes λ_0^{-2} and thereby takes account of spectral information of the forcing. To obtain a version of Leray's inequality in terms of $\kappa_{N,r}^2$, we can also easily see that dividing (6.5.10) by F_0 and time averaging gives

$$\left\langle \kappa_{1,0}^2 \right\rangle \leq \lambda_0^{-2}, \tag{6.5.12}$$

which is true for both $2d$ and $3d$. These results are listed in Tables 6.2 and 6.3.

6.5.3 *Estimates for* $\overline{\lim}_{t\to\infty} F_1$, $\langle F_2 \rangle$, *and* $\left\langle \kappa_{2,1}^2 \right\rangle$

Now we turn to the evolution of the enstrophy. The evolution of the enstrophy $\int_\Omega |\omega|^2 \, d^d x$, is given by[1]

$$\begin{aligned} \frac{1}{2}\frac{d}{dt}\int_\Omega |\omega|^2 \, d^d x &= -\nu \int_\Omega |\nabla\omega|^2 \, d^d x - \int_\Omega \omega \cdot [\mathbf{u}\cdot\nabla\omega] \, d^d x \\ &+ \int_\Omega \omega \cdot [\omega\cdot\nabla\mathbf{u}] \, d^d x + \int_\Omega \omega \cdot \operatorname{curl}\mathbf{f} \, d^d x. \end{aligned} \tag{6.5.13}$$

As we also saw in Chapter 3, because of the periodicity of the boundary conditions and the fact that $\nabla \cdot \mathbf{u} = 0$, the integral $\int_\Omega \omega \cdot [\mathbf{u}\cdot\nabla\omega] \, d^d x$ vanishes for both $2d$ and $3d$. In $2d$, $\omega \cdot \nabla\mathbf{u} = 0$ because the vorticity vector ω is perpendicular to a $2d$ flow in the $x - y$ plane. This enables us to easily find an absorbing ball for the $2d$ enstrophy. In terms of F_1, we find

$$\frac{1}{2}\dot{F}_1 \leq -\nu F_2 + \nu \lambda_0^{-2} F_1. \tag{6.5.14}$$

From Lemma 6.3 we know that

$$F_1^2 \leq F_0 F_2, \tag{6.5.15}$$

and so, using (6.5.9), we find that

$$\overline{\lim}_{t\to\infty} F_1 \leq c\, \lambda_0^{-2} \nu^2 \mathcal{G}^2. \tag{6.5.16}$$

Considering (6.5.14) again, we may time average the equation to obtain $\langle F_2 \rangle$ or we can divide by F_1 and then time average to find $\left\langle \kappa_{2,1}^2 \right\rangle$. In both cases the results appear in Table 6.2 below. The estimate for F_1 in (6.5.16) (showing up also in $\langle F_2 \rangle$ and $\left\langle \kappa_{2,1}^2 \right\rangle$ both being a priori bounded above) is one of the fundamental differences between $2d$ and $3d$ flows. As we shall see in section 7.2, the bound in (6.5.16) will allow us to control the

[1] The enstrophy is defined in section 3.2.

Table 6.2. *A summary of 2d Navier-Stokes estimates*

2*d* Grashof number	$\mathcal{G} = L^2\nu^{-2}\|\mathbf{f}\|_2$
Absorbing ball for F_0	$\overline{\lim}_{t\to\infty} F_0 \le c\,\nu^2\mathcal{G}^2$
Absorbing ball for F_1	$\overline{\lim}_{t\to\infty} F_1 \le c\,\lambda_0^{-2}\nu^2\mathcal{G}^2$
Time average of F_1	$\langle F_1\rangle \le \lambda_0^{-2}\nu^2\mathcal{G}^2$
Time average of F_2	$\langle F_2\rangle \le \lambda_0^{-4}\nu^2\mathcal{G}^2$
Time average of $\kappa_{1,0}^2$	$\langle\kappa_{1,0}^2\rangle \le \lambda_0^{-2}$
Time average of $\kappa_{2,1}^2$	$\langle\kappa_{2,1}^2\rangle \le \lambda_0^{-2}$

whole ladder in the 2*d* case. To date, no such control has been achieved for the 3*d* case.

For a general 3*d* flow, the vortex stretching term $\boldsymbol{\omega}\cdot\nabla\mathbf{u}$ is nonzero. This means that we need to estimate the integral

$$\int_\Omega \boldsymbol{\omega}\cdot[\boldsymbol{\omega}\cdot\nabla\mathbf{u}]\; d^dx. \tag{6.5.17}$$

It is not difficult to see that we can bound it above in the following way

$$\left|\int_\Omega \boldsymbol{\omega}\cdot[\boldsymbol{\omega}\cdot\nabla\mathbf{u}]\; d^3x\right| \le \|\boldsymbol{\omega}\|_\infty\,\|\boldsymbol{\omega}\|_2^2. \tag{6.5.18}$$

Expressing the enstrophy in terms of F_1, we find, in the F_N notation,

$$\frac{1}{2}\dot{F}_1 \le -\nu F_2 + (c\,\|\boldsymbol{\omega}\|_\infty + \nu\lambda_0^{-2})F_1. \tag{6.5.19}$$

A division of (6.5.19) by F_1 and a time averaging operation gives

$$\left\langle\kappa_{2,1}^2\right\rangle \le c\,\nu^{-1}\left\langle\|\boldsymbol{\omega}\|_\infty\right\rangle + \lambda_0^{-2}. \tag{6.5.20}$$

The extra $\|\boldsymbol{\omega}\|_\infty$ term in (6.5.20), in comparison to the 2*d* case in Table 6.2 makes a very great difference. Indeed, with the methods used here, it is mathematically *the* great difference between the two cases. This extra term is the one that has so far stood in the way of finding an acceptable regularity proof for the 3*d* Navier-Stokes equations. The results we have achieved so far in 3*d* are given in Table 6.3.

6.6 A ladder for the thermal convection equations

To illustrate the generality of the method we now turn to the equations for thermal convection which were derived in Chapter 2 for a unit temperature difference across a unit height gap. These have been given

Table 6.3. *A summary of 3d Navier-Stokes estimates*

3*d* Grashof number	$\mathcal{G} = L^{3/2}\nu^{-2}\|\mathbf{f}\|_2$
Absorbing ball for F_0	$\overline{\lim}_{t\to\infty} F_0 \le c\,L\nu^2\mathcal{G}^2$
Time average of F_1	$\langle F_1\rangle \le c\,L\lambda_0^{-2}\nu^2\mathcal{G}^2$ (Leray)
Time average of $\kappa_{1,0}^2$	$\langle\kappa_{1,0}^2\rangle \le \lambda_0^{-2}$
Time average of $\kappa_{2,1}^2$	$\langle\kappa_{2,1}^2\rangle \le c\,\nu^{-1}\langle\|\omega\|_\infty\rangle + \lambda_0^{-2}$

in (2.1.31)-(2.1.35). It is usual in this type of problem to take the temperature T in such a way that it has a linear vertical profile with a variation $\theta(x, y, x, t)$ around it so $T = 1 - z + \theta$. Periodic boundary conditions may then be imposed on $\mathbf{u}$ and θ. The Boussinesq equations are

$$\frac{\partial \mathbf{u}}{\partial t} + \mathbf{u}\cdot\nabla\mathbf{u} + \nabla p = \sigma\Delta\mathbf{u} + \hat{\mathbf{k}}\sigma Ra\,\theta, \tag{6.6.1}$$

$$\nabla\cdot\mathbf{u} = 0, \tag{6.6.2}$$

$$\frac{\partial \theta}{\partial t} + \mathbf{u}\cdot\nabla\theta = \Delta\theta + u_3, \tag{6.6.3}$$

where the dimensionless Prandtl number is defined as

$$\sigma = \frac{\nu}{\kappa}, \tag{6.6.4}$$

and the Rayleigh number is

$$Ra = \frac{\alpha g\delta T h^3}{\rho\nu\kappa}. \tag{6.6.5}$$

The temperature difference δT is now the driving force – there are no body forces as in the previous sections of this chapter. There are few differences between this case and the case with body forces dealt with in previous sections. Let

$$\Theta_N = \int_\Omega |D^N\theta|^2\,d^dx. \tag{6.6.6}$$

Now we combine this with the H_N defined for the velocity field in (6.1.2) and add in a Ra^N term to take account of the forcing

$$\mathcal{F}_N = H_N + \Theta_N + Ra^N. \tag{6.6.7}$$

Referring to (6.2.25) when considering (6.6.1), we obtain

$$\frac{1}{2}\dot{H}_N \le -\sigma H_{N+1} + c\,\|D\mathbf{u}\|_\infty H_N + \sigma Ra\,H_N^{1/2}\theta_N^{1/2}, \tag{6.6.8}$$

where we have used Cauchy's inequality to obtain the last term. Now, when we consider the temperature field θ in (6.6.3), the nonlinear term $\mathbf{u}\cdot\nabla T$ is mixed. We can appeal to Lemma 6.1 to obtain

$$\frac{1}{2}\dot{\Theta}_N \le -\Theta_{N+1} + c\,\Theta_N^{1/2+1/q} H_N^{1/p} \|\mathbf{Du}\|_\infty^{1-a} \|D\theta\|_\infty^{1-b} + H_N^{1/2}\Theta_N^{1/2}, \quad (6.6.9)$$

where, as in Lemma 6.1, a and b are defined in (6.2.19) and (6.2.21). In fact, both these equations imply that $a+b=1$ and $1/p+1/q=1/2$. Combining (6.6.8) and (6.6.9), we obtain

$$\begin{aligned}\frac{1}{2}\dot{\mathscr{F}}_N \le\ & -\sigma H_{N+1} - \Theta_{N+1} - Ra^{N+1} \\ & + c\left(\|\mathbf{Du}\|_\infty + \|D\theta\|_\infty + (\sigma+1)Ra + 1\right)\mathscr{F}_N. \end{aligned} \quad (6.6.10)$$

Finally, using Lemma 6.3, we find we have proved

Theorem 6.3 *For $d=2,3$, the $\mathscr{F}_N$ defined in (6.6.7) evolve according to*

$$\frac{1}{2}\dot{\mathscr{F}}_N \le -min[\sigma,1]\frac{\mathscr{F}_N^{1+1/s}}{\mathscr{F}_{N-s}^{1/s}} + c\left(\|\mathbf{Du}\|_\infty + \|D\theta\|_\infty + (\sigma+1)Ra + 1\right)\mathscr{F}_N. \quad (6.6.11)$$

□

6.7 References and further reading

A chapter on the property of the D operator can be found in the book by John [25]. The ladder theorem as presented here in can be found in [26] and [27]. See also the work of [28]. The ladder rolls together into one simple package, results which have been well known for many years: For a history and further references see Constantin and Foias [12] and Temam [24]. See also Saffman's book [29] for a discussion of vorticity.

Exercises

1 For a general vector field $\mathbf{v}$ defined on the periodic domain Ω in two or three dimensions, show that if $S=\nabla\cdot\mathbf{v}$ and $\omega=\text{curl}\,\mathbf{v}$, then

$$\int_\Omega |\nabla\mathbf{v}|^2\, d^dx = \int_\Omega \left(|\omega|^2 + S^2\right) d^dx, \quad (E6.1)$$

and

$$\int_\Omega |\Delta \mathbf{v}|^2 \, d^d x = \int_\Omega \left(|\mathrm{curl}\, \omega|^2 + |\nabla S|^2 \right) \, d^d x. \tag{E6.2}$$

2 It is possible to formally "solve" the F_N ladder by integration. By defining $Y_N = F_N^{-1/s}$, show that the ladder in equality (6.2.5) in Theorem 6.1 can be rewritten as

$$\dot{Y}_N + A(t) Y_N \geq \frac{2}{s} Y_{N-s}, \tag{E6.3}$$

where

$$A(t) = c \, \|D\mathbf{u}\|_\infty + \nu \lambda_0^{-2}. \tag{E6.4}$$

Hence show that the ladder inequality is equivalent to

$$Y_N I(t) \geq \frac{2}{s} \int_0^t Y_{N-s}(\zeta) I(\zeta) \, d\zeta, \tag{E6.5}$$

where

$$I(t) = \exp \int_0^t A(\tau) \, d\tau. \tag{E6.6}$$

3 There is an alternative version of the H_N and F_N ladders which depend on $\|\mathbf{u}\|_\infty^2$ instead of $\|D\mathbf{u}\|_\infty$. To obtain this, first show that an alternative to either (6.2.12) or (6.2.25) can be found by using the vector identity (E1.1) to obtain the Navier-Stokes equations in the form

$$\mathbf{u}_t + \omega \times \mathbf{u} = \nu \Delta \mathbf{u} - \nabla (p + u^2/2) + \mathbf{f}. \tag{E6.7}$$

Using this and Lemma 6.2, show that

$$|\text{Nonlinear term}| \leq c_N \, H_N^{1/2} H_{N+1}^{1/2} \, \|\mathbf{u}\|_\infty. \tag{E6.8}$$

Hence show that the alternative full ladder is given by

$$\frac{1}{2} \dot{F}_N \leq -\frac{\nu}{2} \frac{F_N^{1+1/s}}{F_{N-s}^{1/s}} + \left(c_N \, \nu^{-1} \, \|\mathbf{u}\|_\infty^2 + \nu \lambda_0^{-2} \right) F_N. \tag{E6.9}$$

4 (The ladder in full norm form) Let

$$M_N = \sum_{j=0}^{N-s} L^{-2j} F_{N-j}. \tag{E6.10}$$

The definition in (E6.10) has coefficients which are powers of L which keep each term the same dimension. In full, M_N can be written

$$M_N = F_N + L^{-2}F_{N-1} + \ldots + L^{-2(N-s)}F_s. \tag{E6.11}$$

For $1 \le s \le N$, use Lemma 6.3 to show that M_N satisfies

$$\frac{1}{2}\dot{M}_N \le -\nu \frac{M_N^{1+1/s}}{M_{N-s}^{1/s}} + \left(c_N \|\mathbf{Du}\|_\infty + \nu\lambda_0^{-2}\right) M_N. \tag{E6.12}$$

Hence, high norms depend on low norms: M_{N-s} can be written

$$M_{N-s} = F_{N-s} + L^{-2}F_{N-s-1} + \ldots + L^{-2(N-s)}F_0. \tag{E6.13}$$

5 Show that for $r < N_1 < N_2$,

$$\kappa_{N_1,r} \le \kappa_{N_2,r}. \tag{E6.14}$$

Also show that $r_1 < r_2 < N$,

$$\kappa_{N,r_1} \le \kappa_{N,r_2}. \tag{E6.15}$$

7

Regularity and length scales for the 2d and 3d Navier-Stokes equations

7.1 Introduction

This chapter continues the theme of Chapter 6 and uses the two main theorems (Theorems 6.1 and 6.2) proved there and the estimates displayed in tables 6.1, 6.2 and 6.3 to discuss (1) regularity in both $2d$ and $3d$ and (2) how estimates for natural small length scales might be computed in both these cases.

For an infinite dimensional dynamical system in a box of side L and spatial dimension d, the problem of resolving the typical length scale ℓ of natural features in the flow is normally related to the number of degrees of freedom $\mathcal{N}$ in the system by

$$\mathcal{N} \sim \left(\frac{L}{\ell}\right)^d . \tag{7.1.1}$$

It desirable to associate the number of degrees of freedom with the average number of data required to configure all possible motions starting from all initial states. In the descriptive way it has been introduced in (7.1.1), $\mathcal{N}$ has not been properly defined. To do this we need firstly to show that a global attractor $\mathcal{A}$ exists for the system in question. Secondly it is necessary to devise a way of associating $\mathcal{N}$ with a typical length scale (or scales) of solutions on this attractor. In fact, $\mathcal{N}$ is often loosely identified with the Lyapunov dimension $d_L(\mathcal{A})$ of the global attractor, but that calculation is left to Chapter 9. It will be seen, however, in future sections of this Chapter that certain estimates for the $2d$ Navier-Stokes equations coincide with some found in that Chapter.

We end this introduction with a caveat: In the $3d$ case it must be borne in mind that because there is yet no proof that strong solutions exist, it must be assumed that one or another time average; such as, for example, $\langle \|D\mathbf{u}\|_\infty \rangle$, is finite and can be manipulated as if strong solutions exist and make sense. The predictions made about minimum length scales are made under this assumption. Unless one can elevate weak to strong solutions, we technically have no a priori estimates; nevertheless, it is useful to do this anyway to see what it is that needs to be proved.

7.2 A global attractor and length scales in the $2d$ case

The ladder theorem expressed in Theorem 6.1, and summarized in Table 6.1, can now be used to see if it is possible to prove that there is an absorbing ball for *each* of the F_N. The main piece of extra information that we can use in the $2d$ case is that we have an absorbing ball for F_1 (see Table 6.2). The task, therefore, is to control the $\|D\mathbf{u}\|_\infty$ term. We use a $2d$ calculus inequality from the Appendix

$$\|D\mathbf{u}\|_\infty \le c\,\|D^N\mathbf{u}\|_2^a\,\|D\mathbf{u}\|_2^{1-a}, \tag{7.2.1}$$

where $a = (N-1)^{-1}$ and $N \ge 3$. In turn this gives

$$\|D\mathbf{u}\|_\infty \le c\,F_N^{a/2}\,F_1^{(1-a)/2}. \tag{7.2.2}$$

Choosing $s = N-1$ in the ladder, we obtain

$$\frac{1}{2}\dot{F}_N \le -\nu\,\frac{F_N^{1+\frac{1}{N-1}}}{F_1^{\frac{1}{N-1}}} + \left(c_N\,F_N^{a/2}F_1^{(1-a)/2} + \nu\lambda_0^{-2}\right)F_N. \tag{7.2.3}$$

It is clear from (7.2.3) that the exponent of the negative term is stronger than that of the nonlinear term enabling us to find an absorbing ball. The result of this is

$$\overline{\lim}_{t\to\infty}F_N \le c_N\,\nu^{-2(N-1)}\left(\overline{\lim}_{t\to\infty}F_1\right)^N + \lambda_0^{-2(N-1)}\left(\overline{\lim}_{t\to\infty}F_1\right). \tag{7.2.4}$$

Now, from Table 6.2, we have an estimate for $\overline{\lim}_{t\to\infty}F_1$ in terms of the $2d$ Grashof number $\mathscr{G}$ which, when substituted into (7.2.4), gives

$$\overline{\lim}_{t\to\infty}F_N \le c\,\lambda_0^{-2N}\nu^2\left(\mathscr{G}^{2N} + \mathscr{G}^2\right). \tag{7.2.5}$$

In in the $2d$ case in section 5.4 we showed how, in (5.4.40), the limit of the Galerkin approximations $\|\omega^N\|_2^2$ gives finite enstrophy. The weak solutions of Chapter 5 can therefore be turned into strong solutions, with all the consequences for uniqueness discussed there. Consequently the bound on F_1 implies the bound on F_N given in (7.2.5) for every $N \ge 1$, so no singularities in any derivative can develop from smooth initial data in a $2d$ flow. The triumph is only minor, however, because $2d$ flows are not particularly physical and their inability to amplify vorticity, due to the absence of the vortex stretching term, makes them a poor substitute for $3d$ flows. As we shall see in future sections, it is the question of control over F_1 around which the difference between the $2d$ and $3d$ cases revolves. These results allow us, however, to consider the idea of a global attractor and length scales of solutions.

7.2.1 A global attractor

The global attractor $\mathscr{A}$ for a finite dimensional dynamical system has been discussed in section 4.2, but this needs to be extended when dealing with infinite dimensional systems. This extension is performed in the following way: If $\mathscr{B}$ is a compact, connected, absorbing set, absorbing all trajectories — the absorbing ball for F_1 — and $S(t)$ is the nonlinear semigroup flow such that $\mathbf{u}(t) = S(t)\mathbf{u}(0)$, then the global attractor $\mathscr{A}$ is defined as

$$\mathscr{A} = \cap_{t>0} S(t)\mathscr{B}. \tag{7.2.6}$$

$\mathscr{A}$ has the properties:

- $S(t)\mathscr{A} = \mathscr{A}$ both forward and backward in time,
- For the ω-limit set of any bounded set $J \in \mathscr{A}$ then $\omega(J) \in \mathscr{A}$,
- $\mathscr{A}$ is compact and has the property

$$\mathrm{dist}_{t\to\infty}\{\mathbf{u}(t), \mathscr{A}\} = 0. \tag{7.2.7}$$

The meaning of the above properties are self-evident taken in context with the discussions in Chapters 4, 5 and 6. From (7.2.5) it also appears that for the $2d$ Navier-Stokes equations, solutions on $\mathscr{A}$ are smooth. In Chapter 8 we go even a step better than this and show that solutions are actually analytic. How to estimate $d_L(\mathscr{A})$ is left to Chapter 9.

7.2.2 Length scales in the 2d Navier-Stokes equations

Now we turn our attention to defining and estimating $\mathscr{N}$. First we use the objects

$$\kappa_{N,r}(t) = \left(\frac{F_N}{F_r}\right)^{\frac{1}{2(N-r)}} \tag{7.2.8}$$

introduced in (6.3.8). These can be thought of as the natural time dependent mean wavenumbers for the system as they all have a dimension of (length)$^{-1}$. They are ordered such that $\kappa_{N,r} \le \kappa_{N+1,r}$ and $\kappa_{N,r} \le \kappa_{N,r+1}$ for $r < N$. Moreover, the smallest $\kappa_{1,0}$ is bounded from below so all of them are bounded away from zero. Secondly we define the dimensionless quantities

$$\mathcal{N}_{N,r} = \lambda_0^2 \, \kappa_{N,r}^2, \tag{7.2.9}$$

where $\lambda_0^{-2} = L^{-2} + \lambda_f^{-2}$ and where normally, $\lambda_f \le L$. From now on we therefore consider λ_0 to be the smallest length scale on the domain. The set of time averages

$$\langle \mathcal{N}_{N,r} \rangle \sim \left(\frac{\lambda_0}{\ell_{N,r}} \right)^2 \tag{7.2.10}$$

can play the role of $\mathcal{N}$ in (7.1.1) for $d = 2$, although it actually has a doubly infinite number of members. λ_0 replaces L and $\ell_{N,r}^{-2} \sim \langle \kappa_{N,r}^2 \rangle$. To see how this is useful, firstly we take the ladder in the form

$$\frac{1}{2}\dot{F}_N \le -\nu F_{N+1} + \left(c_{N,r} \|\mathbf{Du}\|_\infty + \nu \lambda_0^{-2} \right) F_N \tag{7.2.11}$$

from (6.2.43) in Chapter 6. It is necessary to convert this differential inequality in F_N and F_{N+1} into one in $\kappa_{N,r}$ and $\kappa_{N+1,r}$ in a slightly different way from that given in (6.4.5). Taking $s = N - r$ and differentiating (7.2.8) with respect to t, we find that

$$2s\dot{\kappa}_{N,r}\kappa_{N,r}^{2s-1} = \frac{\dot{F}_N}{F_r} - \frac{\dot{F}_r}{F_r}\kappa_{N,r}^{2s}, \tag{7.2.12}$$

(7.2.11) can be used for $\dot{F}_N$ but for $\dot{F}_r$ a lower bound is necessary. In fact, this is available from (6.2.6))

$$\frac{1}{2}\dot{F}_r \ge -F_{r+1} - \left(c_{N,r} \|\mathbf{Du}\|_\infty + \nu \lambda_0^{-2} \right) F_r. \tag{7.2.13}$$

In combination we obtain

$$\begin{aligned} (N-r)\dot{\kappa}_{N,r} \;\le\; & -\nu\kappa_{N+1,r}^3 + \nu\kappa_{r+1,r}^2\kappa_{N,r} \\ & + \; 2\left(c_{N,r} \|\mathbf{Du}\|_\infty + \nu \lambda_0^{-2} \right) \kappa_{N,r}. \end{aligned} \tag{7.2.14}$$

In order to estimate the term $\|\mathbf{Du}\|_\infty$, we first prove the following lemma for a general $2d$ scalar function $A(\mathbf{x})$ and then follow it with a corollary:

Lemma 7.1 *In 2d, the* $\|A\|_\infty$ *norm of a scalar function is bounded as*

$$\|A\|_\infty \le c\,\|DA\|_2 \left[1 + \log\left(L \frac{\|D^2 A\|_2}{\|DA\|_2} \right) \right]^{1/2}. \tag{7.2.15}$$

Corollary 7.1 *In 2d,*

$$\|\mathbf{Du}\|_\infty \le c\,F_2^{1/2} \left[1 + \log\left(\lambda_0 \kappa_{3,2} \right) \right]^{1/2}. \tag{7.2.16}$$

Proof: The proof begins by considering the Fourier transform $\hat{A}(\mathbf{k})$ of $A(\mathbf{x})$ and splitting the spectrum into low modes $|\mathbf{k}| \leq \kappa$ and high modes $|\mathbf{k}| \geq \kappa$. The division point κ will be determined later.

$$\begin{aligned} |A(\mathbf{x})| &= \left|\sum_{\mathbf{k}} \exp(i\mathbf{k}\cdot\mathbf{x})\hat{A}(\mathbf{k})\right| \leq \sum_{|\mathbf{k}|\leq\kappa} |\hat{A}(\mathbf{k})| + \sum_{|\mathbf{k}|\geq\kappa} |\hat{A}(\mathbf{k})| \\ &= \sum_{|\mathbf{k}|\leq\kappa} |\mathbf{k}|^{-1}|\mathbf{k}||\hat{A}(\mathbf{k})| + \sum_{|\mathbf{k}|\geq\kappa} |\mathbf{k}|^{-2}|\mathbf{k}|^{2}|\hat{A}(\mathbf{k})| \\ &\leq \left(\sum_{|\mathbf{k}|\leq\kappa} |\mathbf{k}|^{-2}\right)^{1/2} \left(\sum_{|\mathbf{k}|\leq\kappa} |\mathbf{k}|^{2}|\hat{A}(\mathbf{k})|^{2}\right)^{1/2} \\ &\quad + \left(\sum_{|\mathbf{k}|\geq\kappa} |\mathbf{k}|^{-4}\right)^{1/2} \left(\sum_{|\mathbf{k}|\geq\kappa} |\mathbf{k}|^{4}|\hat{A}(\mathbf{k})|^{2}\right)^{1/2}, \end{aligned} \tag{7.2.17}$$

where we have used the Cauchy-Schwarz inequality in the last step. The leading factor in the first term is estimated as

$$\sum_{|\mathbf{k}|\leq\kappa} |\mathbf{k}|^{-2} \leq cL^2 \int_{2\pi/L}^{\kappa} dk/k \leq cL^2 \log L\kappa, \tag{7.2.18}$$

where $2\pi/L$ is the smallest k in the box. The leading factor in the second term in (7.2.17) obeys

$$\sum_{|\mathbf{k}|\geq\kappa} |\mathbf{k}|^{-4} \leq cL^2 \int_{\kappa}^{\infty} dk/k^3 \leq cL^2\kappa^{-2}. \tag{7.2.19}$$

The second factors in the two terms in (7.2.17) are bounded by the seminorms $\|DA\|_2$ and $\|D^2A\|_2$

$$\sum_{|\mathbf{k}|\leq\kappa} |\mathbf{k}|^2\, |\hat{A}(\mathbf{k})|^2 \leq cL^{-2}\|DA\|_2^2 \tag{7.2.20}$$

and

$$\sum_{|\mathbf{k}|\geq\kappa} |\mathbf{k}|^4\, |\hat{A}(\mathbf{k})|^2 \leq cL^{-2}\|D^2A\|_2^2. \tag{7.2.21}$$

Inserting (7.2.18)-(7.2.21) into (7.2.17) gives

$$\|A\|_\infty \leq c\, \|DA\|_2\, (\log L\kappa)^{1/2} + \frac{\|D^2A\|_2}{\kappa}. \tag{7.2.22}$$

In the above, $\kappa \geq 2\pi/L$ is essentially arbitrary and we may choose

$$\kappa = \frac{\|D^2A\|_2}{\|DA\|_2} \tag{7.2.23}$$

yielding

$$\|A\|_\infty^2 \le c\,\|DA\|_2^2\left[1+\log\left(L\frac{\|D^2A\|_2^2}{\|DA\|_2^2}\right)\right]. \qquad (7.2.24)$$

This completes the proof. □

Proof of corollary: Choose a set of scalar functions A_i such that $A_i = Du_i$ for every component of **u**. Then we return to (7.2.22) and, noting that $\|D^2u_i\|_2^2 \le F_2$ and $\|D^3u_i\|_2^2 \le F_3$, rewrite it as

$$\|\mathbf{Du}\|_\infty \le c\,(\log\lambda_0\kappa)^{1/2}F_2^{1/2} + \frac{F_3^{1/2}}{\kappa}, \qquad (7.2.25)$$

having chosen $2\pi/\lambda_0$ instead of $2\pi/L$ as the lowest mode. Then we choose κ as

$$\kappa^2 = \frac{F_3}{F_2}, \qquad (7.2.26)$$

so κ^2 can be identified as $\kappa_{3,2}^2 = F_3/F_2$. This leaves us with the final result

$$\|\mathbf{Du}\|_\infty \le c\,F_2^{1/2}\left[1+\log\left(\lambda_0\kappa_{3,2}\right)\right]^{1/2}, \qquad (7.2.27)$$

which is the end of the proof. □

Now we return to (7.2.14) and choose $r = 1$

$$\begin{aligned}(N-1)\dot{\kappa}_{N,1} \;&\le\; -\nu\kappa_{N+1,1}^3 + \nu\kappa_{N,1}\kappa_{2,1}^2 + 2\nu\lambda_0^{-2}\kappa_{N,1} \\ &+\; c_3\,F_2^{1/2}\left[1+\log\left(\lambda_0\kappa_{3,2}\right)\right]^{1/2}\kappa_{N,1}. \qquad (7.2.28)\end{aligned}$$

Because every F_N is bounded from below, pointwise in time estimates for $\kappa_{N,r}$ can be found from (7.2.5) (see the exercises)

$$\overline{\lim}_{t\to\infty}\kappa_{N,r} \le \lambda_0^{-1}\mathscr{G}^{\frac{N-1}{N-r}}. \qquad (7.2.29)$$

This is important as use of this in the logarithmic term when $r = 2$ and in the $\kappa_{2,1}^2\kappa_{N,1}$ term when $r = 1$ removes them from contention in future inequalities. Using these, in terms of the $\mathscr{N}_{N,r}$ notation of (7.2.9), a time average of (7.2.28) gives

$$\begin{aligned}\left\langle \mathscr{N}_{N+1,1}^{3/2}\right\rangle \;&\le\; c\,\nu^{-1}\lambda_0^2\left\langle F_2^{1/2}\mathscr{N}_{N,1}^{1/2}\right\rangle\left[1+\log\mathscr{G}\right]^{1/2} \\ &+\; 2\left\langle \mathscr{N}_{N,1}^{1/2}\right\rangle + c\,\mathscr{G}\left\langle \mathscr{N}_{2,1}\right\rangle \qquad (7.2.30)\end{aligned}$$

Invoking the Cauchy-Schwarz inequality and using the estimate for $\langle F_2 \rangle$ from table 6.2, we obtain a recursion relation for $N \geq 2$

$$\begin{aligned} \langle \mathcal{N}_{N+1,1} \rangle^{3/2} &\leq c\mathcal{G}(1+\log\mathcal{G})^{1/2} \langle \mathcal{N}_{N,1} \rangle^{1/2} \\ &+ 2\langle \mathcal{N}_{N,1} \rangle^{1/2} + c\mathcal{G}\langle \mathcal{N}_{2,1} \rangle . \end{aligned} \tag{7.2.31}$$

Because $\langle \kappa_{2,1}^2 \rangle \leq \lambda_0^{-2}$, we have $\langle \mathcal{N}_{2,1} \rangle \leq 1$ as our bottom rung. For $N = 2$, absorbing the last two terms in (7.2.31) into the constant (because $\mathcal{G} > 1$) we obtain

$$\langle \mathcal{N}_{3,1} \rangle = \lambda_0^2 \langle \kappa_{3,1}^2 \rangle \leq c\mathcal{G}^{2/3}(1+\log\mathcal{G})^{1/3}. \tag{7.2.32}$$

Actually, this estimate for $\langle \mathcal{N}_{3,1} \rangle$ coincides with that for the Lyapunov dimension of the attractor that will be derived in Chapter 9. Using the recursion relation (7.2.31) we find that the first few $\langle \mathcal{N}_{N,1} \rangle$ are

$$\langle \mathcal{N}_{2,1} \rangle = \lambda_0^2 \langle \kappa_{2,1}^2 \rangle \leq 1, \tag{7.2.33}$$

$$\langle \mathcal{N}_{3,1} \rangle = \lambda_0^2 \langle \kappa_{3,1}^2 \rangle \leq c_8\, \mathcal{G}^{2/3}(1+\log\mathcal{G})^{1/3}, \tag{7.2.34}$$

$$\langle \mathcal{N}_{4,1} \rangle = \lambda_0^2 \langle \kappa_{4,1}^2 \rangle \leq c_9\, \mathcal{G}^{8/9}(1+\log\mathcal{G})^{4/9}. \tag{7.2.35}$$

The result for $N \gg 1$ is clearly

$$\langle \mathcal{N}_{N,1} \rangle = \lambda_0^2 \langle \kappa_{N,1}^2 \rangle \leq c_{10}\, \mathcal{G}(1+\log\mathcal{G})^{1/2}. \tag{7.2.36}$$

For values of $r \geq 2$, direct estimation through (6.4.3) together with the logarithmic estimate (7.2.16) gives (see the exercises)

$$\langle \mathcal{N}_{N,r} \rangle = \lambda_0^2 \langle \kappa_{N,r}^2 \rangle \leq c_{N,r}\, \mathcal{G}(1+\log\mathcal{G})^{1/2}. \tag{7.2.37}$$

In addition to this it is possible to improve slightly upon the pointwise in time estimate given in (7.2.29). This can be performed by turning (7.2.14) into a differential inequality for $\kappa_{N,r}$ after having used the fact that $-\kappa_{N+1,r} \leq -\kappa_{N,r}$. In terms of the $\mathcal{N}_{N,r}$, this result comes out to be

$$\overline{\lim}_{t\to\infty} \mathcal{N}_{N,r} = \overline{\lim}_{t\to\infty} \left\{ \lambda_0^2 \kappa_{N,r}^2 \right\} \leq \tilde{c}_{N,r}\, \mathcal{G}^2. \tag{7.2.38}$$

The result in (7.2.33) is simply an expression of the fact that $\langle \mathcal{N}_{2,1} \rangle$ is capturing a combination of the box length L and the smallest forcing scale λ_f which are both built into λ_0, whereas the estimate for $\langle \mathcal{N}_{3,1} \rangle$ is that which coincides with the estimate for the attractor dimension[1].

[1] In fact, both the estimate for $\langle \mathcal{N}_{3,1} \rangle$ and that for the attractor dimension $d_L(\mathcal{A})$ in Chapter 9 are dependent on $\langle F_2 \rangle$ but on no higher time average.

7.3 3*d* Navier-Stokes regularity?

In chapter 6 we spent some time discussing how the vortex stretching term $\omega \cdot \nabla \mathbf{u}$ makes a dramatic difference between the 2*d* and the 3*d* cases. This term can stretch, tangle and twist vortex lines producing significant dynamics down to very small scales. The 3*d* Navier-Stokes equations are notorious for the absence of any proof which shows them to be regular beyond a finite time. Indeed we could spend much time demonstrating to the reader the many ways that attempted regularity proofs have failed but instead, we shall discuss two cases and then show what can be proved.

7.3.1 Problems with 3d *Navier-Stokes regularity*

Certain time averaged quantities are bounded a priori. Apart from the L^2 norm of the velocity field $\mathbf{u}(\mathbf{x}, t)$, which is bounded above for all t, the first of these is the time integral of F_1, which is Leray's inequality. In terms of a time average this is

$$\langle F_1 \rangle \le \nu^2 L \lambda_0^{-2} \mathcal{G}^2 \tag{7.3.1}$$

where the $\mathcal{G}$ is the 3*d* Grashof number (see Table 6.3). In addition to (7.3.1), it is possible to show that for weak solutions the infinite set $\langle \kappa_{N,1} \rangle$ is bounded for all N and, in addition, $\langle \|\mathbf{u}\|_\infty \rangle$ is also bounded. We leave these to a subsequent subsection.

Leray's inequality in (7.3.1) is a time average and not a pointwise bound so F_1 can still exhibit singularities in time. Is it possible to find an absorbing ball for F_1 as in the 2*d* case? Let us try and obtain a differential inequality for F_1 by going direct to the equation for the vorticity

$$\frac{1}{2}\dot{F}_1 \le -\nu F_2 + \int_\Omega \omega \cdot (\omega \cdot \nabla)\mathbf{u}\, dV + \nu \lambda_0^{-2} F_1. \tag{7.3.2}$$

In the 2*d* case, the middle term on the right hand side would be normally absent but here, we integrate by parts once, pull out $\|\mathbf{u}\|_\infty$, break it up using the Cauchy-Schwarz inequality and obtain

$$\frac{1}{2}\dot{F}_1 \le -\nu F_2 + \|\mathbf{u}\|_\infty F_1^{1/2} F_2^{1/2} + \nu \lambda_0^{-2} F_1. \tag{7.3.3}$$

Using the calculus inequality (see Appendix (A.0.18)),

$$\|\mathbf{u}\|_\infty \le c\, F_2^{1/4}\, F_1^{1/4}. \tag{7.3.4}$$

and then a Hölder inequality in (7.3.3), we obtain

$$\frac{1}{2}\dot{F}_1 \le -\frac{\nu}{4} F_2 + c\, \nu^{-3} F_1^3 + \nu \lambda_0^{-2} F_1. \tag{7.3.5}$$

The standard use of $F_1^2 \le F_0 F_2$ finally produces

$$\frac{1}{2}\dot{F}_1 \le -\frac{\nu}{4}\frac{F_1^2}{F_0} + c\,\nu^{-3}F_1^3 + \nu\lambda_0^{-2}F_1. \qquad (7.3.6)$$

It is clear that the cube in the nonlinear term is too large for the square in the viscosity term thereby preventing us finding an absorbing ball for arbitarily large initial data for fixed ν. The two exceptions are the special cases of small initial data on F_1 or ν large enough so that the negative term dominates.

The interesting point about (7.3.6) is that the $\nu^{-3}F_1^3$ term is sharp in the following sense. First note that the dimension of every term in the inequality is L^3T^{-3} and if we were to ask whether it is possible to perform inequalities in an alternative way to produce an exponent of F_1 lower than 3, this term would still have to have the same dimension. If one tries to obtain an estimate in terms of the combination $\nu^\alpha F_0^\beta F_1^\gamma$ then it is easy to show that the lowest value of γ that can be achieved with $\beta \ge 0$ is $\gamma = 3$ with $\alpha = -3,\ \beta = 0$.

Let us now look at the problem a different way and examine the $\|D\mathbf{u}\|_\infty$ term in the F_N ladder. A calculus inequality from the Appendix yields

$$\|D\mathbf{u}\|_\infty \le c\,F_N^{a/2}\,\|\mathbf{u}\|_2^{1-a}. \qquad (7.3.7)$$

where $a = \frac{5}{2N}$ and with $2N > 5$. The ladder now becomes

$$\frac{1}{2}\dot{F}_N \le -\nu\frac{F_N^{1+1/s}}{F_{N-s}^{1/s}} + c\,F_N^{1+a/2}\|\mathbf{u}\|_2^{1-a} + \nu\lambda_0^{-2}F_N, \qquad (7.3.8)$$

from which it can be seen that a necessary condition for an absorbing ball is

$$1/s > a/2 \qquad (7.3.9)$$

but where the F_{N-s} term in the denominator needs to be controlled. We should like to be able to go down the ladder as far as $s = N$ because F_0 is bounded above for all t whereas no proof exists that F_1 is bounded in $3d$, as it is in the $2d$ case. Unfortunately, it is clear that an absorbing ball cannot be achieved this way, because equation (7.3.9) means that we need $s < \frac{4N}{5}$: This cannot be fulfilled by taking $s = N$. Therefore, we can only get down to $s = N - 1$ without violating the condition for an absorbing ball. In consequence, F_1 is again the bottom rung of the ladder, as in the $2d$ case, and not F_0. Since we have no control over this bottom rung in $3d$, there is no means of controlling the other rungs of the ladder either. Herein lies the root of the problem. There are several different angles of attack but they all reduce to the same result in the end.

Failure to prove regularity by these methods drives us to see if we can deduce the weakest *assumption* that needs to be made to obtain regularity (see the reference section for a history of this). The idea is to relax the requirement that we go down to $\|\mathbf{u}\|_2^2$ in the calculus inequality in (7.3.7) above and instead go down to L^q (for some q to be calculated) and not L^2. Hence, instead of (7.3.7), we write

$$\|D\mathbf{u}\|_\infty \leq c\, F_N^{a/2} \|\mathbf{u}\|_q^{1-a}, \tag{7.3.10}$$

where $a = \frac{2(q+3)}{(2N-3)q+6}$ with $N \geq 3$. The requirement in (7.3.9) that s must satisfy $as < 2$ to get an absorbing ball means that there is the following restriction on q

$$q > \frac{3(N-2)}{N-3}. \tag{7.3.11}$$

However large N is taken, it is clear that q must always satisfy $q > 3$. In conclusion therefore, if we assume that $\|\mathbf{u}\|_q$ $(q > 3)$ is bounded for all t, then all the F_N are bounded for all t and no singularities can occur. However, no independent proof exists that shows that $\|\mathbf{u}\|_{3+\varepsilon}$ is a priori bounded. The barrier between what we have proved is bounded ($\|\mathbf{u}\|_2$) and what we need to prove is bounded ($\|\mathbf{u}\|_{3+\varepsilon}$) therefore appears small, but has so far remained insurmountable.

7.3.2 *A Bound on* $\langle \kappa_{N,1} \rangle$ *in* 3*d*

Can anything be said about the objects $\kappa_{N,r}$ in $3d$? Unlike the $2d$ case of the previous section, no regularity proof is known so a finite time singularity in any of the $\kappa_{N,r}$ cannot be ruled out. Consequently, all the following manipulations are formal and the results are valid only for weak solutions. Nevertheless, we will show that there are bounds on the infinite set of time averaged quantities $\langle \kappa_{N,1} \rangle$. The first step is to use (7.2.14) with $r = 1$ and use $-\kappa_{N+1,1}^3 \leq -\kappa_{N,1}^3$.

$$(N-1)\dot{\kappa}_{N,1} \leq -\nu\kappa_{N,1}^3 + \nu\kappa_{2,1}^2\kappa_{N,1} + 2\left(c_{N,1}\|D\mathbf{u}\|_\infty + \nu\lambda_0^{-2}\right)\kappa_{N,1}. \tag{7.3.12}$$

Recall the definition of λ_K, the Kolmogorov length, first given in (3.3.18)

$$\lambda_K = \left(\frac{\nu^3}{\varepsilon}\right)^{1/4} \tag{7.3.13}$$

where the energy dissipation[1] rate, an a priori bounded quantity for weak solutions, is defined by $\varepsilon = \nu \langle F_1 \rangle L^{-3}$

Lemma 7.2 *In 3d, weak solutions of the Navier-Stokes equations satisfy*

$$\langle \kappa_{2,1} \rangle \leq c\, L\lambda_0^{-2}\mathscr{G}^2, \tag{7.3.14}$$

or, alternatively,

$$L \langle \kappa_{2,1} \rangle \leq c \left(\frac{L}{\lambda_K} \right)^4, \tag{7.3.15}$$

Proof: In $3d$, we already have an inequality for F_1 in the shape of (7.3.5), namely

$$\frac{1}{2}\dot{F}_1 \leq -\frac{\nu}{4}F_2 + c\,\nu^{-3}F_1^3 + \nu\lambda_0^{-2}F_1. \tag{7.3.16}$$

Now divide by F_1^2 and time average to get

$$\left\langle \frac{F_2}{F_1^2} \right\rangle \leq c\,\nu^{-4}\langle F_1 \rangle \leq c\, L\lambda_0^{-2}\nu^{-2}\mathscr{G}^2. \tag{7.3.17}$$

We have ignored a correction term proportional to $\langle F_1^{-1} \rangle$ which is small and bounded because F_1 is bounded from below. Rewriting (7.3.17) in terms of the Kolmogorov length λ_K, we find that

$$\left\langle \frac{F_2}{F_1^2} \right\rangle \leq c\,\nu^{-4}\langle F_1 \rangle \leq c\,\nu^{-2}\frac{L^3}{\lambda_K^4}. \tag{7.3.18}$$

Now $\langle \kappa_{2,1} \rangle$ can be written as

$$\begin{aligned} \langle \kappa_{2,1} \rangle &= \left\langle \left(\frac{F_2}{F_1} \right)^{1/2} \right\rangle = \left\langle \frac{F_2^{1/2}}{F_1} F_1^{1/2} \right\rangle \\ &\leq \left\langle \frac{F_2}{F_1^2} \right\rangle^{1/2} \langle F_1 \rangle^{1/2} \\ &\leq c\, L\lambda_0^{-2}\mathscr{G}^2 \end{aligned} \tag{7.3.19}$$

which is (7.3.14). The alternative upper bound in terms of λ_K follows. This ends the proof.

This result enables us to prove the following:

[1] Sometimes we use $\langle H_1 \rangle$ and sometimes $\langle F_1 \rangle$ for convenience, the only difference being the constant forcing terms.

Theorem 7.4 *For $N \geq 2$, weak solutions of the 3d Navier-Stokes equations satisfy*

$$\langle \kappa_{N,1} \rangle \leq c_N \, L \lambda_0^{-2} \mathcal{G}^2, \tag{7.3.20}$$

or, alternatively,

$$L \langle \kappa_{N,1} \rangle \leq c_N \left(\frac{L}{\lambda_K} \right)^4. \tag{7.3.21}$$

Proof: To handle the $\|D\mathbf{u}\|_\infty$ term we use a calculus inequality from the Appendix

$$\begin{aligned} \|D\mathbf{u}\|_\infty &\leq c \, \|D^N \mathbf{u}\|_2^a \, \|D\mathbf{u}\|_2^{1-a} \leq c \, F_N^{a/2} \, F_1^{(1-a)/2} \\ &= c \, \kappa_{N,1}^{3/2} \, F_1^{1/2}, \end{aligned} \tag{7.3.22}$$

for $N \geq 3$ where $a = \frac{3}{2(N-1)}$. Now we divide (7.3.12) through by $\kappa_{N,1}^2$ and use the fact that the $\kappa_{N,1}$ are bounded below and that $\kappa_{2,1} \leq \kappa_{N,1}$ for $N \geq 2$ to get

$$\langle \kappa_{N,1} \rangle \leq c \, \nu^{-1} \left\langle \kappa_{N,1}^{1/2} F_1^{1/2} \right\rangle + \langle \kappa_{2,1} \rangle + 2\lambda_0^{-2} \kappa_{N,1}^{-1}. \tag{7.3.23}$$

Splitting up the first term on the right hand side of (7.3.23) and using Hölder's inequality we find

$$\langle \kappa_{N,1} \rangle \leq c_N \, \nu^{-2} \, \langle F_1 \rangle + 2 \, \langle \kappa_{2,1} \rangle, \qquad N \geq 3, \tag{7.3.24}$$

where we have ignored the last term in (7.3.23) which is bounded and small. Use of Lemma 7.2 allows us to control $\langle \kappa_{2,1} \rangle$ which then gives the advertised result. The alternative upper bound (7.3.21) in terms of λ_K follows as in Lemma 7.2. □

7.3.3 Bounds on $\langle \|\mathbf{u}\|_\infty \rangle$ and $\left\langle \|D\mathbf{u}\|_\infty^{1/2} \right\rangle$

Following from Theorem 7.4, there are two more quantities that can be bounded from above. The first follows from the identity

$$\left\langle F_N^{\frac{1}{2N-1}} \right\rangle = \left\langle \kappa_{N,r}^{\frac{2(N-r)}{2N-1}} \, F_r^{\frac{1}{2N-1}} \right\rangle. \tag{7.3.25}$$

Using a Hölder inequality, this becomes

$$\left\langle F_N^{\frac{1}{2N-1}} \right\rangle \leq \langle \kappa_{N,r} \rangle^{\frac{2(N-r)}{2N-1}} \left\langle F_r^{\frac{1}{2r-1}} \right\rangle^{\frac{2r-1}{2N-1}}. \tag{7.3.26}$$

Starting with $r = 1$ and then using the fact that both $\langle F_1 \rangle$ and $\langle \kappa_{N,1} \rangle$ are bounded above, we have

$$\left\langle F_N^{\frac{1}{2N-1}} \right\rangle \le c_N \, \nu^{\frac{2}{2N-1}} L \lambda_0^{-2} \mathcal{G}^2, \tag{7.3.27}$$

or

$$\left\langle F_N^{\frac{1}{2N-1}} \right\rangle \le c_N \, \nu^{\frac{2}{2N-1}} L^3 \lambda_K^{-4}. \tag{7.3.28}$$

Secondly, an inequality for $\langle \|\mathbf{u}\|_\infty \rangle$ comes by using a higher order version of (7.3.4) (see exercises) which can be expressed in terms of $\kappa_{N,1}$,

$$\|\mathbf{u}\|_\infty \le c\, \kappa_{N,1}^{1/2} F_1^{1/2}. \tag{7.3.29}$$

Using the Cauchy-Schwarz inequality, we find that

$$\langle \|\mathbf{u}\|_\infty \rangle \le c\, \nu L \lambda_0^{-2} \mathcal{G}^2. \tag{7.3.30}$$

The method here not only makes plain the inter-relation between the triad of objects $\langle \kappa_{N,1} \rangle$, $\langle F_N^{\frac{1}{2N-1}} \rangle$ and $\langle \|\mathbf{u}\|_\infty \rangle$, it also shows that controlling $\langle \kappa_{N,1} \rangle$ is the keystone to all these results.

What of $\langle \|D\mathbf{u}\|_\infty \rangle$? The best we can do with these methods is to estimate $\left\langle \|D\mathbf{u}\|_\infty^{1/2} \right\rangle$. Using (7.3.22) we find that

$$\left\langle \|D\mathbf{u}\|_\infty^{1/2} \right\rangle \le c \left\langle \kappa_{N,1}^{3/4} F_1^{1/4} \right\rangle \le c \, \langle \kappa_{N,1} \rangle^{3/4} \langle F_1 \rangle^{1/4}. \tag{7.3.31}$$

A combination of (7.3.21) and (7.3.13) gives

$$\left\langle \|D\mathbf{u}\|_\infty^{1/2} \right\rangle \le c\, \nu^{1/2} L^3 \lambda_K^{-4}. \tag{7.3.32}$$

This will be of use in Chapter 9 when estimating the attractor dimension in the $3d$ case.

7.4 The Kolmogorov length and intermittency

Because no proof exists which shows that a $3d$ flow remains regular beyond a finite time, any predictions about the flow and its natural length scales must be based on an assumption that one of the various norms involved remains sufficiently regular for time averages to make sense. We are particularly interested in the possibility that even if the flow remains regular, the occurrence of large, rare, intermittent fluctuations, or bursts in the vorticity field away from averages, may nevertheless act as a source of small scale structures in the flow. If this is the case, a flow may remain close to its spatial average for large periods of time but such bursts in the vorticity field might cause excursions of energy to short scales. The deeper the excursions, the rarer these events must be to avoid violating the bounds on space and time averages. This phenomenon, called *intermittency*, is associated with energy dissipation being concentrated on a set of dimension lower than the background space. Because of the relatively long quiescent periods between these events, one might contemplate how to compute a length scale associated with the flow during these periods, and then a second smaller scale associated with a burst, if or when it occurs. It is in this situation that we should like to investigate the role of the Kolmogorov length λ_K, which was introduced in Chapter 3 in (3.3.18). In particular, we should like to see whether λ_K is the smallest scale, as is commonly assumed, or whether length scales smaller than this could possibly occur.

The estimate from Theorem (6.2) for $\left\langle \kappa_{N,r}^2 \right\rangle$, namely

$$\left\langle \kappa_{N,r}^2 \right\rangle \leq c\,\nu^{-1} \left\langle \|D\mathbf{u}\|_\infty \right\rangle + \lambda_0^{-2}, \tag{7.4.1}$$

is distinguished by the fact that it is formally an intensive quantity, i.e., it not explicitly dependent on the system size. We could define a "Kolmogorov length" from this by defining an "energy dissipation rate" $\varepsilon_\infty = \nu \left\langle \|D\mathbf{u}\|_\infty \right\rangle^2$. Dropping the suffices on the scales, this gives

$$\ell_{N,r}^{-2} \leq c\,\lambda_{K,\infty}^{-2} + \lambda_0^{-2} \tag{7.4.2}$$

with

$$\lambda_{K,\infty} = \left(\frac{\nu^3}{\varepsilon_\infty} \right)^{1/4}. \tag{7.4.3}$$

Although $\lambda_{K,\infty}$ is *mathematically* a length scale, it is not the quantity which is normally understood to be the Kolmogorov length. How are we to make a comparison between $\lambda_{K,\infty}$ and λ_K? In so doing, one runs into a difficult problem. Using the calculus inequalities of the Appendix to bound above the L^∞ norm of a function by the L^2-norm of its Nth derivative (say) introduces the system volume into the problem, thereby sacrificing the intuitive notion of an intensive length scale determined only by local properties of the flow. Let us begin by writing down how many fixed (or bounded) scales exist against which we want to compare $\nu^{-1} \left\langle \|D\mathbf{u}\|_\infty \right\rangle$.

- $\lambda_0^{-2} = L^{-2} + \lambda_f^{-2}$; the box length L and the smallest forcing scale λ_f.
- The conventional Kolmogorov length $\lambda_K = (\nu^3/\varepsilon)^{1/4}$.

- The alternative Kolmogorov length

$$\Lambda_K = \left(\frac{\nu^3}{\bar{\varepsilon}}\right)^{1/4}, \tag{7.4.4}$$

defined with an energy dissipation rate $\bar{\varepsilon} = \nu L^{-3} \sup_t F_1$. While we have no absolute certainty that $\bar{\varepsilon}$ is bounded, we assume here that it is. We return to this idea again in Chapter 9 in section 9.3.

There are now two alternative approximations that can be pursued. The first is to we make the assumption

$$\langle \|\boldsymbol{Du}\|_\infty \rangle \approx L^{-3/2} \langle \|\boldsymbol{Du}\|_2 \rangle, \tag{7.4.5}$$

this makes

$$\ell_{N,r}^{-2} \leq c\,\lambda_{K,\infty}^{-2} + \lambda_0^{-2} \approx c\,\lambda_K^{-2} + \lambda_0^{-2}. \tag{7.4.6}$$

Physically, this assumption means that the dissipation in the flow is close to its spatial average in the mean square sense most of the time, and that any large deviations do not contribute significantly to the time average. Strong intermittency would certainly violate the assumption (7.4.5) in a fundamental way. In addition, making this approximation in (7.3.12) gives the result

$$\left\langle \kappa_{N,1}^3 \right\rangle \leq c_N \lambda_K^{-3}, \qquad N \geq 2, \tag{7.4.7}$$

thereby making the estimate for the number of degrees of freedom defined in (7.1.1) to be

$$\mathcal{N} \leq c_N \left(\frac{L}{\lambda_K}\right)^3. \tag{7.4.8}$$

This uniformly makes λ_K the average small scale when fluctuations are ignored. The second way of pursuing this idea in order to take into account fluctuations away from $\|\boldsymbol{Du}\|_2$ is not to approximate $\|\boldsymbol{Du}\|_\infty$ but to estimate it in terms of $\|\boldsymbol{Du}\|_2$. Using the calculus inequalities from (7.3.22) we obtain

$$\left\langle \kappa_{N,1}^2 \right\rangle \leq c\,\nu^{-1} \left\langle \kappa_{N,1}^{3/2} \|\boldsymbol{Du}\|_2 \right\rangle + \lambda_0^{-2}. \tag{7.4.9}$$

Using a Hölder inequality we obtain

$$\left\langle \kappa_{N,1}^2 \right\rangle \leq c\,\nu^{-4} \left\langle F_1^2 \right\rangle + 4\lambda_0^{-2}. \tag{7.4.10}$$

Now the $\nu^{-4}\langle F_1^2 \rangle$ term has no known bound but if we use the Kolmogorov length Λ_K defined in (7.4.4) we can pull one F_1 outside the time average to get

$$\frac{\ell}{L} \geq c \left(\frac{\lambda_K}{L}\right)^2 \left(\frac{\Lambda_K}{L}\right)^2 \tag{7.4.11}$$

where $\ell^{-2} \sim \langle \kappa_{N,1}^2 \rangle$. This estimate for ℓ is certainly not intensive because of its explicit L dependence. Because $\lambda_K^{-1} \leq \Lambda_K^{-1}$, (7.4.11) comes out as

$$\frac{\ell}{L} \geq c \left(\frac{\Lambda_K}{L}\right)^4. \tag{7.4.12}$$

This form of small scale estimate will appear again in Chapter 8 in (8.4.16).

7.5 Singularities and the Euler equations

An important issue in the theory of inviscid fluid dynamics is whether the 3d Euler equations

$$\mathbf{u}_t + \mathbf{u} \cdot \nabla \mathbf{u} = -\nabla p, \tag{7.5.1}$$

with $\nabla \cdot \mathbf{u} = 0$, can generate a finite time singularity from smooth initial data. Certain results concerning the 3d Euler equations can be demonstrated using the methods developed here. Consider the results for the evolution of $\kappa_{N,r}$ for the Navier-Stokes equations from Chapter 6, expressed in both row 6 of Table 6.1 in section 6.5, and also in (7.3.12), but with $\nu = 0$ and $\mathbf{f} = 0$

$$(N-r)\dot{\kappa}_{N,r} \leq c_{N,d} \, \|D\mathbf{u}\|_\infty \kappa_{N,r}. \tag{7.5.2}$$

This is valid for the Euler equations with the forcing terms removed from the F_N. A simple integration with respect to time gives

$$\kappa_{N,r}(t) \leq \kappa_{N,r}(0) \exp\left[c_N \int_0^t \|D\mathbf{u}\|_\infty(\tau) d\tau\right]. \tag{7.5.3}$$

In addition, because $\kappa_{N,r} \leq \kappa_{N+1,r}$ (see the exercises of Chapter 6), (7.5.2) is also valid in the case of 3d Navier-Stokes, with an additional $\left(\nu\lambda_0^{-2}\right) \kappa_{N,r}$ term. (7.5.3) shows that $\int_0^t \|D\mathbf{u}\|_\infty(\tau) d\tau$ controls any singularities that might form. For the Euler equations a more natural question to ask is whether $\int_0^t \|\omega\|_\infty(\tau)\, d\tau$ controls possible singularity formation. We now state a theorem which was first proved by Beale, Kato, and Majda (see reference section), valid for the periodic domain $[0, L]^3$:

Theorem 7.5

$$\|D\mathbf{u}\|_\infty \le c\|\omega\|_\infty \left[1+\log^+(L\kappa_{N,r})\right] + L^{-3/2}\|\omega\|_2 \qquad (7.5.4)$$

for $N \ge 3$ *and* $0 \le r < N$. *The* $+$ *sign on the logarithm is defined by* $\log^+ a = \log a$ *for* $a \ge 1$ *and* $\log^+ a = 0$ *otherwise.*

Proof: Let us now consider solutions for the velocity field $\mathbf{u}(\mathbf{x},t)$ through the Biot-Savart law which inverts $\omega = \operatorname{curl}\mathbf{u}$

$$\mathbf{u}(\mathbf{x}) = \int_\Omega \mathbf{K}(\mathbf{x}-\xi)\times\omega(\xi)\,d^3\xi, \qquad (7.5.5)$$

where $\mathbf{K}$ is the gradient of the kernel of the inverse of $-\Delta$ on mean zero periodic functions on $[0,L]^3$. We introduce a cut-off function $\zeta_\rho(\mathbf{x})$, satisfying $\zeta_\rho(\mathbf{x}) = 1$ for $|\mathbf{x}| < \rho$, $\zeta_\rho(\mathbf{x}) = 0$ for $|\mathbf{x}| > 2\rho$, and $|D\zeta_\rho(\mathbf{x})| \le c/\rho$. Here ρ $(0 < \rho \le L/4)$ is a radius chosen suitably small later. We introduce a factor $\zeta_\rho(\mathbf{x}-\xi) + [1-\zeta_\rho(\mathbf{x}-\xi)]$ under the integral sign and split $D\mathbf{u}$ into two terms, the first being

$$D\mathbf{u}^{(1)} = \int \zeta_\rho(\mathbf{x}-\xi)\mathbf{K}(\mathbf{x}-\xi)\times D\omega(\xi)\,d^3\xi. \qquad (7.5.6)$$

Furthermore, the coincidence point singularity in $\mathbf{K}$ is bounded according to $|\mathbf{K}(\mathbf{x}-\xi)| \le c\,|\mathbf{x}-\xi|^{-2}$, so by Hölder's inequality

$$\|D\mathbf{u}^{(1)}\|_\infty \le \|K\|_p\|D\omega\|_q \le c\rho^{1-3/q}\|D\omega\|_q \qquad (7.5.7)$$

with the restriction $q > 3$ where $1/p + 1/q = 1$. Hence $p < 3/2$. Now we use a calculus inequality

$$\|D\omega\|_q \le c\|D^{N-1}\omega\|_2^a\|\omega\|_2^{1-a} \qquad (7.5.8)$$

where, with $N \ge 3$,

$$a = \frac{3}{N-1}\left(\frac{5}{6}-\frac{1}{q}\right), \qquad (7.5.9)$$

and so

$$\|D\mathbf{u}^{(1)}\|_\infty^2 \le c\rho^{2(1-3/q)}\kappa_{N,1}^{2(N-1)a}F_1. \qquad (7.5.10)$$

After a little manipulation, we are left with the second term

$$D\mathbf{u}^{(2)} = \int D\left[\left(1-\zeta_\rho(\mathbf{x}-\xi)\right)\mathbf{K}(\mathbf{x}-\xi)\right]\times\omega(\xi)\,d^3\xi, \qquad (7.5.11)$$

where the integral is over $\rho \leq |\mathbf{x}-\xi| \leq L$. For $D\mathbf{u}^{(2)}$, we estimate the two terms in the gradient separately and use $|D\mathbf{K}| \leq |\mathbf{x}-\xi|^{-3}$ to obtain

$$\|D\mathbf{u}^{(2)}\|_\infty \leq c\left(\int_\rho^L r^{-3}r^2 dr + \int_\rho^{2\rho} r^{-2}\rho^{-1}r^2 dr\right)\|\omega\|_\infty, \tag{7.5.12}$$

which finally gives

$$\|D\mathbf{u}\|_\infty \leq c\rho^{1-3/q}\kappa_{N,1}^{\frac{5}{2}-\frac{3}{q}}F_1^{1/2} + c\|\omega\|_\infty\left[1+\log\left(\frac{L}{\rho}\right)\right]. \tag{7.5.13}$$

Now we choose

$$\rho^{-1} = c\kappa_{N,1}^{\frac{5q-6}{2(q-3)}} L^{\frac{3q}{2(q-3)}} \tag{7.5.14}$$

which, in (7.5.13) gives the result of the theorem. □

The main point here is that the $\kappa_{N,r}$ $(N \geq 3)$ are the natural mediators between $\|D\mathbf{u}\|_\infty$ and $\|\omega\|_\infty$. Another consequence of Theorem 7.5 is that we can estimate the $\kappa_{N,r}$ explicitly as functions of time in terms of the time integral of $\|\omega\|_\infty$. An integration of (7.5.2) using Theorem 7.5 gives

$$L\kappa_{N,r}(t) \leq c\,\exp\left(\int_0^t g(\tau)\exp[I(t)-I(\tau)]\,d\tau\right), \tag{7.5.15}$$

where

$$\begin{aligned} I(t) &= \int_0^t \|\omega\|_\infty(\tau)\,d\tau, \\ g(t) &= L^{-3/2}\|\omega\|_2. \end{aligned} \tag{7.5.16}$$

This means that no singularity can develop at t^* without the quantity $\int_0^{t^*}\|\omega\|_\infty(\tau)\,d\tau \to \infty$ at t^* also. This result guarantees that a solution of the $3d$ Euler equations cannot develop a singularity through the following processes:

- The development of kinks or curvature singularities in vortex lines.
- The development of "vorticity shocks" where ω becomes discontinuous but $\|\omega\|_\infty$ remains bounded.
- The rate of strain matrix $S_{ij} = \frac{1}{2}\left(u_{i,j}+u_{j,i}\right)$ becomes singular while the vorticity remains bounded.

Consequently, if any one of these types of behavior appears in a numerically computed solution, then it must be an artifact of the numerical scheme. Furthermore, it shows that $\int_0^t\|\omega\|_\infty(\tau)\,d\tau$ *alone* controls any singularities that might form. It is not sufficient, for instance, to monitor $\|\omega\|_2$ as this may still remain bounded while $\|\omega\|_\infty \to \infty$.

If a numerical integration scheme indicates that a singularity of the type $\|\omega\|_\infty \sim (t^* - t)^{-\gamma}$, forms at t^* then the BKM theorem says that $\int_0^{t^*} \|\omega\|_\infty(\tau)\,d\tau$ must also become singular at t^*. Hence we must have $\gamma \geq 1$ for the singularity to be genuine and not an artefact of the numerics.

7.6 References and further reading

After Leray's work in the 1930s [23], the work of Ladyzhenskaya [30] and Serrin [31] on regularity of the $3d$ Navier-Stokes equations should be studied: This post-1960 approach to both the $2d$ and $3d$ cases has an extensive literature associated with it: In addition to Constantin and Foias [12] and Temam [14, 24], see Lions and Magenes [32]. For work on global attractors see Hale [33] and Vishik [34] in addition to [12] and [14]. The infinite set of time averaged quantities $\langle F_N^{\frac{1}{2N-1}} \rangle$ was first published by Foias, Guillopé, and Temam in [35]. The approach displayed in sections 7.2 and 7.3 for estimating length scales appears in [36] which is based on earlier work in [26]. The paper by Beale, Kato, and Majda [37] contains the original proof of Theroem 7.5 concerning the Euler equations in section 7.4 which shows how $\int_0^t \|\omega\|_\infty(\tau)\,d\tau$ controls singularity formation. Theorem 7.5 is a reworking of their original proof: Estimates for the kernel of the inverse Laplacian can be found, for instance, in Glimm and Jaffe [38]. The two reviews by Majda [39, 40] contain many ideas and references on modern aspects of the interaction between analysis and numerical methods in fluid mechanics.

Exercises

1 Use the calculus inequality $\|D\mathbf{u}\|_\infty \leq c\,F_N^{a/2}\|\mathbf{u}\|_q^{1-a}$ for $q > 1$ in the ladder theorem for the $3d$ Navier-Stokes equations to show that an absorbing ball for F_N can be found only if $q > 3$.

2 For the $2d$ Navier-Stokes equations, use the ladder for $\kappa_{N,r}$ given in Table 6.1 and Theorem 7.1 to verify (7.2.29), namely that

$$\overline{\lim}_{t\to\infty}\kappa_{N,r} \leq \lambda_0^{-1}\mathscr{G}^{\frac{N-1}{N-r}}, \tag{E7.1}$$

by considering λ_0 as the smallest scale on the domain. Show also that an estimate which is pointwise in t exists in the form

$$\kappa_{N,2}(t) \leq c\,\kappa_{N,2}(0)\exp\left(t\lambda_0^{-2}\nu\mathscr{G}\right). \tag{E7.2}$$

3 Verify the estimate given in (7.2.36) in the $2d$ case, namely

$$\langle \mathcal{N}_{N,r} \rangle = \lambda_0^2 \left\langle \kappa_{N,r}^2 \right\rangle \le c_{N,r}\, \mathcal{G}(1+\log \mathcal{G})^{1/2} \tag{E7.3}$$

by using Corollary 7.1.

4 For the $3d$ Euler equations show that

$$\frac{d}{dt}\int_\Omega \mathbf{u}\cdot\boldsymbol{\omega}\, d^3x = 0, \tag{E7.4}$$

when $\boldsymbol{\omega}\cdot\mathbf{n} = 0$ on $\partial\Omega$. The quantity $\int_\Omega \mathbf{u}\cdot\boldsymbol{\omega}\, d^3x$ is called the helicity.

5 Prove the inequality from the Appendix in (A.0.18) for $N \ge 2$, namely

$$\|f\|_\infty^2 \le c\,\|\Delta f\|_2\,\|\nabla f\|_2. \tag{E7.5}$$

Hint: While this can be done by Fourier analysis an alternative is to try two calculus inequalities in combination

$$\|f\|_\infty \le c\,\|\Delta f\|_2^a\,\|\nabla f\|_q^{1-a}, \tag{E7.6}$$

and

$$\|f\|_q \le c\,\|\nabla f\|_2^b\,\|f\|_\infty^{1-b}. \tag{E7.7}$$

6 Use (7.3.5) to show that in $3d$

$$\left\langle \kappa_{2,1}^2 \right\rangle \le c\,\nu^{-4}\left\langle F_1^2 \right\rangle + \lambda_0^{-2}. \tag{E7.8}$$

7 Show that

$$\left\langle \|\mathbf{u}\|_\infty^2 \right\rangle^{2(N-1)} \le \langle F_N \rangle \left(L\lambda_0^{-2}\nu^2\mathcal{G}^2 \right)^{2N-3} \tag{E7.9}$$

Show also that the result in (7.3.30) also implicitly gives an inverse length scale λ_T^{-1} such that

$$\left(\frac{\lambda_T}{L}\right)^{-1} \le c\left(\frac{L}{\lambda_0}\right)^2 \mathcal{G}^2. \tag{E7.10}$$

8

Exponential decay of the Fourier power spectrum

8.1 Introduction

In this chapter we show how the velocity vector field's entire Fourier mode spectrum is controlled by the energy dissipation rate. Explicit bounds on the amplitudes of the Fourier modes are derived in terms of the magnitude of the L^2 norm of the derivatives of the velocity field. If $\|\nabla \mathbf{u}(\cdot, t)\|_2^2$ is uniformly bounded in time, then the amplitude of the Fourier components $\hat{\mathbf{u}}(\mathbf{k}, t)$ decay exponentially as $|\mathbf{k}| \to \infty$, uniformly in time, after an initial transient. This result establishes that square integrability of the derivatives of the velocity vector field, pointwise in time, is sufficient to ensure that the velocity field is actually infinitely differentiable, a fact that was mentioned, although not proved, in Chapter 5.

In section 8.2 we set up the calculation and derive the relevant nonlinear estimates. The following section, 8.3, is concerned with the solution of the resulting differential inequality. Finally, in 8.4 we show how this allows us to deduce explicit exponentially decreasing bounds on the amplitude of the Fourier modes, and to estimate the length scale associated with this decay of the power spectrum. This length scale, expressed in terms of the energy dissipation rate, is the same as one of the natural scales derived from the ladder structure in Chapters 6 and 7.

8.2 A differential inequality for $\left\|e^{\alpha t|\nabla|}\nabla \mathbf{u}\right\|_2^2$

The specific goal of this section is to derive an evolution inequality for

$$\left\|e^{\alpha t|\nabla|}\nabla \mathbf{u}(\cdot, t)\right\|_2^2 = L^{-d} \sum_{\mathbf{k}} e^{2\alpha t|\mathbf{k}|}\, |\mathbf{k}|^2\, |\hat{\mathbf{u}}(\mathbf{k}, t)|^2, \tag{8.2.1}$$

for solutions of the Navier-Stokes equations on $\Omega = [0, L]^d$ with periodic boundary conditions, and for some $\alpha > 0$, and for some interval of time $[0, t)$.

The operator denoted by the exponential of the absolute value of the derivative is defined in the Fourier transfomed representation: On a square integrable periodic function $f(\mathbf{x})$ with Fourier transform

$$\hat{f}(\mathbf{k}) = \int_\Omega e^{-i\mathbf{k}\cdot\mathbf{x}} f(\mathbf{x})\, d^d x, \tag{8.2.2}$$

the operator $e^{\lambda|\nabla|}$ acts according to

$$\left(e^{\lambda|\nabla|} f\right)(\mathbf{x}) = L^{-d} \sum_{\mathbf{k}} e^{i\mathbf{k}\cdot\mathbf{x} + \lambda|\mathbf{k}|} \hat{f}(\mathbf{k}). \tag{8.2.3}$$

That is, it multiplies the $\mathbf{k}$th Fourier mode by $e^{\lambda|\mathbf{k}|}$.

For f to be in the domain of this operator, $e^{\lambda|\nabla|} f$ must be square integrable, equivalent to the condition that $e^{\lambda|\mathbf{k}|}\hat{f}(\mathbf{k})$ is square summable. When $\lambda > 0$ this means that the amplitude of the Fourier modes must fall off very fast as $|\mathbf{k}| \to \infty$, essentially exponentially. In particular it implies that $|\mathbf{k}|^n \hat{f}(\mathbf{k})$ is square summable for every $n > 0$, so that all derivatives of f are square integrable, and in fact bounded. Functions in the domain of $e^{\lambda|\nabla|}$ are infinitely differentiable. Note that for the periodic boundary conditions considered here, $e^{\lambda|\nabla|}$ commutes with derivatives. It also commutes with the projection operators $\mathbf{P}^N$, onto the first N Fourier modes, as defined in Chapter 5.

The Galerkin approximations defined in Chapter 5, $\mathbf{u}^N = \mathbf{P}^N\{\mathbf{u}\}$, are certainly in the domain of $e^{\lambda|\nabla|}$ because their Fourier components vanish identically for large $|\mathbf{k}|$. We are thus justified in applying $e^{\lambda|\nabla|}$ to these fields, and the following calculations will be made for the Galerkin approximations, although we will not explicitly carry the order of the truncation N. The final estimates will be independent of N, and so will be valid for limits of the $\mathbf{u}^N$ as $N \to \infty$.

We proceed by deriving the exact evolution equation for the square of the L^2 norm of $e^{\alpha t|\nabla|}\nabla\mathbf{u}$. For simplicity we consider the force-free case, although a body force can be included without significant complications. Using the equations of motion (5.3.27) and (5.3.28) for the Galerkin approximations, and noting the commutation of $e^{\alpha t|\nabla|}$ with ∇ and $\mathbf{P}^N$, the usual integrations by parts yield

$$\begin{aligned}\frac{1}{2}\frac{d}{dt}\left\|e^{\alpha t|\nabla|}\nabla\mathbf{u}(\cdot,t)\right\|_2^2 &= \int_\Omega\left(e^{\alpha t|\nabla|}\frac{\partial u_i}{\partial x_j}\right)\left(e^{\alpha t|\nabla|}\frac{\partial^2 u_i}{\partial t\partial x_j}+\alpha e^{\alpha t|\nabla|}|\nabla|\frac{\partial u_i}{\partial x_j}\right)d^dx\\ &= -\nu\left\|e^{\alpha t|\nabla|}\Delta\mathbf{u}\right\|_2^2+\int_\Omega\left(e^{\alpha t|\nabla|}\Delta\mathbf{u}\right)\cdot\left(e^{\alpha t|\nabla|}\mathbf{u}\cdot\nabla\mathbf{u}\right)d^dx\\ &\quad+\alpha\int_\Omega\left(e^{\alpha t|\nabla|}\frac{\partial u_i}{\partial x_j}\right)\left(e^{\alpha t|\nabla|}|\nabla|\frac{\partial u_i}{\partial x_j}\right)d^dx. \qquad (8.2.4)\end{aligned}$$

The action of the operator $|\nabla|$ is defined in terms of the Fourier transform as

$$|\nabla|\mathbf{u}(\mathbf{x},t)=L^{-d}\sum_{\mathbf{k}}e^{i\mathbf{k}\cdot\mathbf{x}}|\mathbf{k}|\hat{\mathbf{u}}(\mathbf{k},t). \qquad (8.2.5)$$

Noting that $|\nabla|\mathbf{u}$ and $\nabla\mathbf{u}$ have the same L^2 norm, the last term may be estimated

$$\begin{aligned}&\alpha\int_\Omega\left(e^{\alpha t|\nabla|}\frac{\partial u_i}{\partial x_j}\right)\left(e^{\alpha t|\nabla|}|\nabla|\frac{\partial u_i}{\partial x_j}\right)d^dx\\ &\leq \alpha\left\|e^{\alpha t|\nabla|}\nabla\mathbf{u}\right\|_2\left\|e^{\alpha t|\nabla|}\Delta\mathbf{u}\right\|_2\\ &\leq \frac{\nu}{2}\left\|e^{\alpha t|\nabla|}\Delta\mathbf{u}\right\|_2^2+\frac{\alpha^2}{2\nu}\left\|e^{\alpha t|\nabla|}\nabla\mathbf{u}\right\|_2^2. \qquad (8.2.6)\end{aligned}$$

Hence (8.2.4) may be rewritten

$$\begin{aligned}\frac{d}{dt}\left\|e^{\alpha t|\nabla|}\nabla\mathbf{u}(\cdot,t)\right\|_2^2 &\leq -\nu\left\|e^{\alpha t|\nabla|}\Delta\mathbf{u}\right\|_2^2+\frac{\alpha^2}{\nu}\left\|e^{\alpha t|\nabla|}\nabla\mathbf{u}\right\|_2^2\\ &\quad+2\int_\Omega\left(e^{\alpha t|\nabla|}\Delta\mathbf{u}\right)\cdot\left(e^{\alpha t|\nabla|}\mathbf{u}\cdot\nabla\mathbf{u}\right)d^dx. \qquad (8.2.7)\end{aligned}$$

It remains to bound the last integral above, arising from the nonlinear term in the dynamical equations, so as to close the differential inequality.

The structure of the last term in (8.2.7) is best seen in the Fourier transformed representation. Direct application of the definitions and a simple integration give

$$\begin{aligned}&\int_\Omega\left(e^{\lambda|\nabla|}\Delta\mathbf{u}\right)\cdot\left(e^{\lambda|\nabla|}\mathbf{u}\cdot\nabla\mathbf{u}\right)d^dx\\ &=-iL^{-2d}\sum_{\mathbf{k}}e^{\lambda|\mathbf{k}|}|\mathbf{k}|^2\hat{\mathbf{u}}(\mathbf{k})^*\cdot\left(\sum_{\mathbf{k}'+\mathbf{k}''=\mathbf{k}}e^{\lambda|\mathbf{k}|}\hat{\mathbf{u}}(\mathbf{k}')\cdot\mathbf{k}''\hat{\mathbf{u}}(\mathbf{k}'')\right). \qquad (8.2.8)\end{aligned}$$

The key to the desired result is the triad nature of the wavenumber interactions generated by the nonlinearity, i.e., the constraint

$$\mathbf{k}'+\mathbf{k}''=\mathbf{k} \qquad (8.2.9)$$

in the inner sum in (8.2.8). The triangle inequality $|\mathbf{k}| \leq |\mathbf{k}'| + |\mathbf{k}''|$ implies that, for $\lambda > 0$

$$e^{\lambda|\mathbf{k}|} \leq e^{\lambda|\mathbf{k}'|}e^{\lambda|\mathbf{k}''|} \tag{8.2.10}$$

so that (8.2.8) can be bounded

$$\begin{aligned} L^{-2d}&\left|\sum_{\mathbf{u}} e^{\lambda|\mathbf{k}|}|\mathbf{k}|^2\hat{\mathbf{u}}(\mathbf{k})^* \cdot \sum_{\mathbf{k}'+\mathbf{k}''=\mathbf{k}} e^{\lambda|\mathbf{k}|}\hat{\mathbf{u}}(\mathbf{k}') \cdot \mathbf{k}''\hat{\mathbf{u}}(\mathbf{k}'')\right| \\ \leq\ & L^{-2d}\sum_{\mathbf{k}} e^{\lambda|\mathbf{k}|}|\mathbf{k}|^2|\hat{\mathbf{u}}(\mathbf{k})| \\ &\times \sum_{\mathbf{k}'+\mathbf{k}''=\mathbf{k}} \left(e^{\lambda|\mathbf{k}'|}|\hat{\mathbf{u}}(\mathbf{k}')|\right)\left(e^{\lambda|\mathbf{k}''|}|\mathbf{k}''||\hat{\mathbf{u}}(\mathbf{k}'')|\right). \end{aligned} \tag{8.2.11}$$

Now define the periodic, mean zero function $w(\mathbf{x})$ by its Fourier transform, $\hat{w}(\mathbf{k})$, as

$$\hat{w}(\mathbf{k}) = e^{\lambda|\mathbf{k}|}|\hat{\mathbf{u}}(\mathbf{k})|, \tag{8.2.12}$$

whereby

$$w(\mathbf{x}) = L^{-d}\sum_{\mathbf{k}} e^{i\mathbf{k}\cdot\mathbf{x}}\hat{w}(\mathbf{k}). \tag{8.2.13}$$

Then the right-hand side of (8.2.11) is

$$\begin{aligned} &L^{-2d}\sum_{\mathbf{k}} e^{\lambda|\mathbf{k}|}|\mathbf{k}|^2|\hat{\mathbf{u}}(\mathbf{k})| \sum_{\mathbf{k}'+\mathbf{k}''=\mathbf{k}} \left(e^{\lambda|\mathbf{k}'|}|\hat{\mathbf{u}}(\mathbf{k}')|\right)\left(e^{\lambda|\mathbf{k}''|}|\mathbf{k}''||\hat{\mathbf{u}}(\mathbf{k}'')|\right) \\ =\ & L^{-2d}\sum_{\mathbf{k}} |\mathbf{k}|^2\hat{w}(\mathbf{k}) \sum_{\mathbf{k}'+\mathbf{k}''=\mathbf{k}} \hat{w}(\mathbf{k}')|\mathbf{k}''|\hat{w}(\mathbf{k}'') \\ =\ & \int_\Omega (\Delta w(\mathbf{x}))\, w(\mathbf{x}) \left(|\nabla|w(\mathbf{x})\right) d^dx. \end{aligned} \tag{8.2.14}$$

In the integral representation this expression can be bounded by pulling out w in L^∞, and Δw and $|\nabla|w$ in L^2:

$$\int_\Omega (\Delta w(\mathbf{x}))\, w(\mathbf{x}) \left(|\nabla|w(\mathbf{x})\right) d^dx \leq \|w\|_\infty \|\nabla w\|_2 \|\Delta w\|_2. \tag{8.2.15}$$

Note that we have used the fact that $|\nabla|w$ has the same L^2 norm as ∇w.

In $2d$ and $3d$, the L^∞ norm of w can be bounded in terms of the L^2 norms of the gradient and the Laplacian of w. We will concentrate on the $3d$ case here, leaving the $2d$ case to the exercises. The relevant estimate in $3d$ is

$$\|w\|_\infty \leq c\, \|\Delta w\|_2^{1/2} \|\nabla w\|_2^{1/2}, \tag{8.2.16}$$

where c is an absolute constant that does not depend on any of the parameters of the problem (see the exercises). This may be proved as follows. Write $w(\mathbf{x})$ as the sum of its Fourier series and bound the sum by the most brutal of estimates,

$$|w(\mathbf{x})| = \left| L^{-3} \sum_{\mathbf{k}} e^{-i\mathbf{k}\cdot\mathbf{x}} \hat{w}(\mathbf{k})\, d^d x \right| \leq L^{-3} \sum_{\mathbf{k}} |\hat{w}(\mathbf{k})|. \tag{8.2.17}$$

Break up the sum into low and high wavenumber components, with the exact location of the division at a wavenumber $\Lambda \geq 2\pi/L$ which is yet to be determined. That is

$$L^{-3} \sum_{\mathbf{k}} |\hat{w}(\mathbf{k})| = L^{-3} \sum_{|\mathbf{k}| \leq \Lambda} |\hat{w}(\mathbf{k})| + L^{-3} \sum_{|\mathbf{k}| > \Lambda} |\hat{w}(\mathbf{k})|. \tag{8.2.18}$$

Estimate the first partial sum in (8.2.18) using Cauchy's inequality,

$$\begin{aligned} L^{-3} \sum_{|\mathbf{k}| \leq \Lambda} |\hat{w}(\mathbf{k})| &= L^{-3} \sum_{|\mathbf{k}| \leq \Lambda} |\mathbf{k}|^{-1} |\mathbf{k}| |\hat{w}(\mathbf{k})| \\ &\leq \left(L^{-3} \sum_{|\mathbf{k}| \leq \Lambda} |\mathbf{k}|^{-2} \right)^{1/2} \left(L^{-3} \sum_{|\mathbf{k}| \leq \Lambda} |\mathbf{k}|^2 |\hat{w}(\mathbf{k})|^2 \right)^{1/2}. \end{aligned} \tag{8.2.19}$$

Note that

$$L^{-3} \sum_{|\mathbf{k}| \leq \Lambda} |\mathbf{k}|^{-2} \leq c \int_{2\pi/L}^{\Lambda} \frac{k^2 dk}{k^2} \leq c\Lambda, \tag{8.2.20}$$

where c is an absolute constant, and

$$L^{-3} \sum_{|\mathbf{k}| \leq \Lambda} |\mathbf{k}|^2 |\hat{w}(\mathbf{k})|^2 \leq \|\nabla w\|_2^2. \tag{8.2.21}$$

In a similar spirit, the second partial sum in (8.2.18) satisfies

$$\begin{aligned} L^{-3} \sum_{|\mathbf{k}| > \Lambda} |\hat{w}(\mathbf{k})| &= L^{-3} \sum_{|\mathbf{k}| > \Lambda} |\mathbf{k}|^{-2} |\mathbf{k}|^2 |\hat{w}(\mathbf{k})| \\ &\leq \left(L^{-3} \sum_{|\mathbf{k}| > \Lambda} |\mathbf{k}|^{-4} \right)^{1/2} \left(L^{-3} \sum_{|\mathbf{k}| > \Lambda} |\mathbf{k}|^4 |\hat{w}(\mathbf{k})|^2 \right)^{1/2}, \end{aligned} \tag{8.2.22}$$

with

$$L^{-3} \sum_{|\mathbf{k}| > \Lambda} |\mathbf{k}|^{-4} \leq c' \int_{\Lambda}^{\infty} \frac{k^2 dk}{k^4} = c'' \Lambda^{-1} \tag{8.2.23}$$

for another absolute constant c'', and

$$L^{-3} \sum_{|\mathbf{k}|>\Lambda} |\mathbf{k}|^4 |\hat{w}(\mathbf{k})|^2 \leq \|\Delta w\|_2^2. \tag{8.2.24}$$

Together with (8.2.18), these give

$$|w(\mathbf{x})| \leq 2\max\{c', c''\} \left(\Lambda^{1/2}\|\nabla w\|_2 + \Lambda^{-1/2}\|\Delta w\|_2\right). \tag{8.2.25}$$

The uniform bound in (8.2.25) is good for any choice of $\Lambda \geq 2\pi/L$, in particular for

$$\Lambda = \frac{\|\Delta w\|_2}{\|\nabla w\|_2}, \tag{8.2.26}$$

which, by Poincaré's inequality, is greater than or equal to $2\pi/L$. This yields the uniform upper estimate

$$|w(\mathbf{x})| \leq 2\max\{c', c''\}\|\nabla w\|_2^{1/2}\|\Delta w\|_2^{1/2}, \tag{8.2.27}$$

from which (8.2.16) follows.

Returning to the problem at hand, the integral in (8.2.15) is seen to obey

$$\int_\Omega (\Delta w(\mathbf{x}))\, w(\mathbf{x})\left(|\nabla| w(\mathbf{x})\right) d^d x \leq c\, \|\nabla w\|_2^{3/2}\|\Delta w\|_2^{3/2}. \tag{8.2.28}$$

The L^2 norms of Δw and ∇w are simply related to the norms of the original quantity of interest, namely $e^{\lambda|\nabla|}\mathbf{u}$. Recalling (8.2.12), we have

$$\begin{aligned} \|\nabla w\|_2^2 &= L^{-d}\sum_{\mathbf{k}} |\mathbf{k}|^2 |\hat{w}(\mathbf{k})|^2 \\ &= L^{-d}\sum_{\mathbf{k}} |\mathbf{k}|^2 \left|e^{\lambda|\mathbf{k}|}\hat{\mathbf{u}}(\mathbf{k})\right|^2 = \left\|e^{\lambda|\nabla|}\nabla\mathbf{u}(\cdot, t)\right\|_2^2, \end{aligned} \tag{8.2.29}$$

and likewise,

$$\begin{aligned} \|\Delta w\|_2^2 &= L^{-d}\sum_{\mathbf{k}} |\mathbf{k}|^4 |\hat{w}(\mathbf{k})|^2 \\ &= L^{-d}\sum_{\mathbf{k}} |\mathbf{k}|^4 \left|e^{\lambda|\mathbf{k}|}\hat{\mathbf{u}}(\mathbf{k})\right|^2 = \left\|e^{\lambda|\nabla|}\Delta\mathbf{u}(\cdot, t)\right\|_2^2. \end{aligned} \tag{8.2.30}$$

Inserting (8.2.29) and (8.2.30) into (8.2.28), and recalling (8.2.14), (8.2.11), and (8.2.8), we have the desired control of the last term in the differential inequality in (8.2.7):

$$2\left|\int_\Omega \left(e^{\alpha t|\nabla|}\Delta\mathbf{u}\right)\cdot\left(e^{\alpha t|\nabla|}\mathbf{u}\cdot\nabla\mathbf{u}\right) d^d x\right| \leq c\left\|e^{\alpha t|\nabla|}\nabla\mathbf{u}\right\|_2^{3/2}\left\|e^{\alpha t|\nabla|}\Delta\mathbf{u}\right\|_2^{3/2}. \tag{8.2.31}$$

Breaking up the right-hand side above using $ab \le a^p/p + b^q/q$, with $p = 4/3$ and $q = 4$, we find

$$2\left| \int_\Omega \left(e^{\alpha t|\nabla|} \Delta \mathbf{u} \right) \cdot \left(e^{\alpha t|\nabla|} \mathbf{u} \cdot \nabla \mathbf{u} \right) d^d x \right| \le c\nu^{-3} \left\| e^{\alpha t|\nabla|} \nabla \mathbf{u} \right\|_2^6 + \nu \left\| e^{\alpha t|\nabla|} \Delta \mathbf{u} \right\|_2^2, \tag{8.2.32}$$

where c has been redefined again. This, then, implies that the differential inequality in (8.2.7) closes to become

$$\frac{d}{dt} \left\| e^{\alpha t|\nabla|} \nabla \mathbf{u}(\cdot, t) \right\|_2^2 \le \alpha^2 \nu^{-1} \left\| e^{\alpha t|\nabla|} \nabla \mathbf{u} \right\|_2^2 + c\nu^{-3} \left\| e^{\alpha t|\nabla|} \nabla \mathbf{u} \right\|_2^6. \tag{8.2.33}$$

We remind the reader that (8.2.33) holds for each of the Galerkin approximations, and that the constant c is absolute in that it depends neither on $\mathbf{u}, \alpha, t, L$, nor N, the order of the Galerkin truncation.

8.3 A bound on $\left\| e^{\alpha t|\nabla|} \nabla \mathbf{u} \right\|_2^2$

Now let

$$z(t) = \left\| e^{\alpha t|\nabla|} \nabla \mathbf{u}(\cdot, t) \right\|_2^2. \tag{8.3.1}$$

The differential inequality in (8.2.33) is

$$\frac{dz}{dt} \le \frac{\alpha^2}{\nu} z + \frac{c}{\nu^3} z^3, \tag{8.3.2}$$

with the initial condition

$$z(0) = \|\nabla \mathbf{u}_0\|_2^2, \tag{8.3.3}$$

where $\mathbf{u}_0(\mathbf{x})$ is the initial velocity vector field under consideration for the Galerkin equations, which is the modal truncation of the initial data under contemplation for the Navier-Stokes equations, presumed to have square-integrable derivatives (i.e., the initial flow field has a finite rate of energy dissipation).

The differential inequality in (8.3.2) may be solved by rewriting it as

$$e^{\alpha^2 t/\nu} \frac{d}{dt} \left(e^{-\alpha^2 t/\nu} z(t) \right) = \frac{dz}{dt} - \frac{\alpha^2}{\nu} z \le \frac{c}{\nu^3} z^3, \tag{8.3.4}$$

leading to

$$\frac{d}{dt} \left(e^{-\alpha^2 t/\nu} z(t) \right) \le \frac{c}{\nu^3} e^{2\alpha^2 t/\nu} \left(e^{-\alpha^2 t/\nu} z(t) \right)^3. \tag{8.3.5}$$

Denoting the natural dependent variable by $y(t)$,

$$y(t) = e^{-\alpha^2 t/\nu} z(t), \tag{8.3.6}$$

(8.3.5) may be manipulated to give the integral relation

$$\int_{y(0)}^{y(t)} \frac{dy}{y^3} \leq \frac{c}{\nu^3} \int_0^t e^{2\alpha^2 t'/\nu} dt'. \tag{8.3.7}$$

After a little elementary algebra we find

$$y(t) \leq \frac{y_0}{\sqrt{1 - c\left(\frac{y_0}{\alpha\nu}\right)^2 \left(e^{2\alpha^2 t/\nu} - 1\right)}}, \tag{8.3.8}$$

with initial condition $y(0) = y_0$. Reintroducing $z(t) = \left\| e^{\alpha t|\nabla|} \nabla \mathbf{u}(\cdot, t) \right\|_2^2$ and noting that $z(0) = y(0) = \|\nabla \mathbf{u}_0\|_2^2$, we conclude

$$\left\| e^{\alpha t|\nabla|} \nabla \mathbf{u}(\cdot, t) \right\|_2^2 \leq \frac{e^{\alpha^2 t/\nu} \|\nabla \mathbf{u}_0\|_2^2}{\sqrt{1 - c\left(\frac{\|\nabla \mathbf{u}_0\|_2^2}{\alpha\nu}\right)^2 \left(e^{2\alpha^2 t/\nu} - 1\right)}}. \tag{8.3.9}$$

This means that $\left\| e^{\alpha t|\nabla|} \nabla \mathbf{u}(\cdot, t) \right\|_2^2$ is finite on the interval $[0, t^*)$, where

$$t^* = \frac{\nu}{2\alpha^2} \log\left(1 + \frac{\alpha^2 \nu^2}{c\, \|\nabla \mathbf{u}_0\|_2^4}\right). \tag{8.3.10}$$

The smaller the initial energy decay rate $\sim \|\nabla \mathbf{u}_0\|_2$ is, the larger t^*. Similarly, the larger the parameter α, the shorter t^*. These considerations make physical sense, because the energy dissipation rate is a measure of how much energy is at the shorter scales and α is related to the proposed exponential decay rate of the energy in the high wavenumber modes. In the next section we pursue this line of thought to bound the amplitudes of the individual modes uniformly in time under the assumption that the energy dissipation rate remains finite.

The bound in (8.3.9) has been derived for the Galerkin truncation $\mathbf{u}^N$. The only place where N appears on the right-hand side is in the initial condition. But because $\|\nabla \mathbf{u}^N\|_2 \leq \|\nabla \mathbf{u}\|_2$, we can take $\|\nabla \mathbf{u}_0\|_2$ to be the L^2 norm of the full, untruncated initial velocity field. Then the upper bound is uniform in N, and so it holds for any limit of the Galerkin approximations, in particular for the weak solutions to the Navier-Stokes equations. Similarly, the time t^* is estimated uniformly by taking $\mathbf{u}_0$ to be the full initial velocity field.

From this point on, then, we can assert that (8.3.9) and (8.3.10) hold for the full Navier-Stokes equations, and drop the reference to the Galerkin approximations.

8.4 Decay of the Fourier spectrum

Each Fourier mode amplitude can be individually controlled from the results of the last section. Indeed, the crudest of estimates gives

$$e^{2\alpha t|\mathbf{k}|}|\mathbf{k}|^2 \, |\hat{\mathbf{u}}(\mathbf{k},t)|^2 \le \sum_{\mathbf{k}} e^{2\alpha t|\mathbf{k}|}|\mathbf{k}|^2 \, |\hat{\mathbf{u}}(\mathbf{k},t)|^2 = L^3 \left\| e^{\alpha t|\nabla|}\nabla\mathbf{u}(\cdot,t)\right\|_2^2. \quad (8.4.1)$$

According to (8.3.9), then the $\mathbf{k}$th mode is bounded as

$$|\hat{\mathbf{u}}(\mathbf{k},t)|^2 \le \frac{L^3}{|\mathbf{k}|^2}\frac{e^{\alpha^2 t/\nu - 2\alpha t|\mathbf{k}|}\,\|\nabla\mathbf{u}_0\|_2^2}{\sqrt{1-c\left(\frac{\|\nabla\mathbf{u}_0\|_2^2}{\alpha\nu}\right)^2\left(e^{2\alpha^2 t/\nu}-1\right)}} \quad (8.4.2)$$

during the interval $[0, t^*)$ where, as before,

$$t^* = \frac{\nu}{2\alpha^2}\log\left(1+\frac{\alpha^2\nu^2}{c\,\|\nabla\mathbf{u}_0\|_2^4}\right). \quad (8.4.3)$$

The upper bound in (8.4.2) is a finite function of t over the interval $[0, t^*)$. It exhibits a local minimum, the location and value of which depends on the specific value of $|\mathbf{k}|$ under consideration, as well as the choice of α. What we will do first is to choose t and then adjust α to produce explicit bounds.

According to (8.4.2), the exponential decay length of the spectrum at time $t < t^*$ is αt. Choose t to be half of t^* and define the associated length

$$\lambda^* = \frac{\alpha t^*}{2} = \frac{\nu}{4\alpha}\log\left(1+\frac{\alpha^2\nu^2}{c\,\|\nabla\mathbf{u}_0\|_2^4}\right). \quad (8.4.4)$$

Note that λ^* is a concave function of $\alpha > 0$, so we may maximize it over the choices of α. The extremizing α is determined by

$$0 = \frac{\partial\lambda^*}{\partial\alpha} = \frac{\nu^3}{4c\|\nabla\mathbf{u}_0\|_2^4}\left\{-\frac{c\|\nabla\mathbf{u}_0\|_2^4}{\nu^2\alpha^2}\log\left(1+\frac{\alpha^2\nu^2}{c\|\nabla\mathbf{u}_0\|_2^4}\right)+\frac{2}{1+\frac{\alpha^2\nu^2}{c\|\nabla\mathbf{u}_0\|_2^4}}\right\}. \quad (8.4.5)$$

This fixes α as

$$\alpha = \frac{\gamma\sqrt{c}\|\nabla\mathbf{u}_0\|_2^2}{\nu}, \quad (8.4.6)$$

where γ^2 is the positive root of

$$0 = -\frac{1}{\gamma^2}\log\left(1+\gamma^2\right)+\frac{2}{1+\gamma^2}. \quad (8.4.7)$$

Hence the "best" length scale associated with the exponential decay at $t = t^*/2$ is

$$\lambda^* = \frac{1}{4\gamma\sqrt{c}} \log\left(1+\gamma^2\right) \frac{\nu^2}{\|\nabla \mathbf{u}_0\|_2^2} = c' \frac{\nu^2}{\|\nabla \mathbf{u}_0\|_2^2}, \tag{8.4.8}$$

where c' is an absolute constant. We conclude that

$$|\hat{\mathbf{u}}(\mathbf{k}, t^*/2)|^2 \leq \frac{e^{-\lambda^*|\mathbf{k}|}}{|\mathbf{k}|^2} \frac{L^3(1+\gamma^2)^{1/4} \|\nabla \mathbf{u}_0\|_2^2}{\sqrt{1-\gamma^{-2}\left[(1+\gamma^2)^{1/4}-1\right]}}. \tag{8.4.9}$$

If we knew that $\|\nabla \mathbf{u}\|_2^2$ was bounded uniformly in time over some interval of time $[0, T)$ where $T > t^*$, that is *if*

$$\sup_{0 \leq t \leq T} \|\nabla \mathbf{u}(\cdot, t)\|_2^2 < \infty, \tag{8.4.10}$$

then the argument above can be repeated to produce a uniform bound on the decay of the Fourier spectrum over the interval $[t^*/2, T + t^*/2)$. Define the uniform length scale in the spirit of (8.4.8),

$$\lambda = c' \frac{\nu^2}{\sup_{0 \leq t \leq T} \|\nabla \mathbf{u}(\cdot, t)\|_2^2}. \tag{8.4.11}$$

Then at each interval of time $t \in [t^*/2, T + t^*/2)$ we can substitute $\|\nabla \mathbf{u}(\cdot, t - t^*/2)\|_2^2$ for $\|\nabla \mathbf{u}_0\|_2^2$ in the results above to obtain the uniform bound

$$|\hat{\mathbf{u}}(\mathbf{k}, t)|^2 \leq c\, L^3 \frac{e^{-\lambda|\mathbf{k}|}}{|\mathbf{k}|^2} \left(\sup_{0 \leq t \leq T} \|\nabla \mathbf{u}(\cdot, t)\|_2^2\right). \tag{8.4.12}$$

The conclusion is that if the energy dissipation rate is bounded uniformly in time, then after a transient time of length $t^*/2$, the Fourier spectrum of the flow field decays exponentially at high wavenumbers. Moreover, the decay length is estimated in terms of the energy dissipation rate as

$$\lambda \sim \frac{\nu^2}{\sup_{0 \leq t \leq T} \|\nabla \mathbf{u}(\cdot, t)\|_2^2} = \frac{\nu^3}{\varepsilon_1 L^3}, \tag{8.4.13}$$

where we have defined the largest instantaneous energy dissipation rate (per unit mass)

$$\varepsilon_1 = L^{-3} \nu \sup_{0 \leq t \leq T} \|\nabla \mathbf{u}(\cdot, t)\|_2^2. \tag{8.4.14}$$

In the conventional scaling theory of turbulence, the Kolmogorov length

λ_{K_1} is the unique length scale constructed from ν and ε_1 without explicit reference to the system size L,

$$\lambda_{K_1} = \left(\frac{\nu^3}{\varepsilon_1}\right)^{1/4}. \qquad (8.4.15)$$

Thus the length scale controlling the exponential decay of the Fourier spectrum of the solutions of the Navier-Stokes equations, derived rigorously here under the assumption that the energy dissipation rate is finite, can be expressed as

$$\frac{\lambda}{L} \sim \left(\frac{\lambda_{K_1}}{L}\right)^4. \qquad (8.4.16)$$

This length scale, naturally interpreted as a lower bound on the "smallest" scale in the flow, is essentially the same as that found in the previous chapter in (7.4.11). In both of these derivations we do not know a priori that this scale λ_{K_1} is nonvanishing, which is precisely the regularity problem for the $3d$ Navier-Stokes equations.[1] The conclusion is that if there was a finite bound on the energy dissipation rate at each instant of time, then we could deduce that the solutions actually had exponentially decaying Fourier mode amplitudes, and were thus infinitely differentiable functions with the energy effectively concentrated at length scales larger than λ. As has been pointed out a number of times already, it is currently not known if this is the case. The results of this chapter highlight the importance of the challenge to produce such estimates.

8.5 References and further reading

The approach developed in this chapter comes directly from Foias and Temam [41]. The length scale associated with the exponential decay of the Fourier spectrum is discussed in Doering and Titi [41].

Exercises

1. Starting from (5.3.27), derive (8.2.4).
2. Compute the constant c in (8.2.16).
3. A slightly improved bound on the Fourier spectrum decay follows from (8.4.2) by first minimizing the right-hand side over t, for each individual $|\mathbf{k}|$, and then minimizing the result over α. Carry out this procedure.

[1] The conventional definition of λ_K involves ε, the time averaged energy dissipation rate, which *is* bounded from below a priori.

4 Mimicking the analysis in this chapter, show that solutions of the 2*d* Navier-Stokes equations on $[0, L]^2$ with periodic boundary conditions have exponentially bounded Fourier mode amplitudes at high wavenumbers. Estimate the small length scales associated with the decay of the power spectrum. In place of (8.2.16), use the 2*d* estimate in (7.2.15) (Lemma 7.1).

5 Rework the analysis in this chapter including a body force $\mathbf{f}(\mathbf{x})$ in the Navier-Stokes equations. Assume that the spectrum of the force has a sharp cutoff at $|\mathbf{k}| = 2\pi/\lambda_f$.

9

The attractor dimension for the Navier-Stokes equations

9.1 Introduction

In this chapter we show how the dimension of the global attractor $\mathscr{A}$ can be estimated for the Navier-Stokes equations. The approach is an extension of that developed in Chapter 4 for ordinary differential equations where it was shown that if N-dimensional volume elements in the system phase space contract to zero, then the attractor dimension $d_L(\mathscr{A})$ must be bounded by N. For partial differential equations the technical chore remains the same; namely, to derive estimates on the spectrum of the linearized evolution operator, linearized around solutions on the attractor, and to perform this operation in some function space instead of an a priori finite dimensional phase space. As we saw in Chapter 4 in the context of the Lorenz equations, this requires some knowledge of the location of the attractor, i.e., a priori estimates on the solutions. This approach is pursued in section 9.2 which deals with the $2d$ Navier-Stokes equations. It was shown in Chapter 7 that a global attractor $\mathscr{A}$ exists in this case, and we have good control of the solutions on the attractor. It turns out that the result for periodic boundary conditions is quite sharp, within logarithms of both the conventional heuristic estimate for the number of degrees of freedom in a $2d$ turbulent flow and rigorous lower bounds.

The $3d$ Navier-Stokes equations on a periodic domain are the concern of section 9.3. The lack of a regularity proof for this case results in some uncertainty concerning the very existence of a compact attractor. To achieve any formal estimate of the attractor dimension it is necessary to assume that H_1 remains bounded for all t. Then we produce estimates for the attractor dimension in terms of a Kolmogorov length Λ_K based on this quantity instead of using the more conventional definition of

λ_K which normally uses $\langle H_1 \rangle$. This estimate turns out to being close to sharp.

9.2 The $2d$ attractor dimension estimate

We begin with the $2d$ forced, incompressible Navier-Stokes equations

$$\frac{\partial \mathbf{u}}{\partial t} + \mathbf{u} \cdot \nabla \mathbf{u} = \nu \Delta \mathbf{u} - \nabla p + \mathbf{f}, \tag{9.2.1}$$

where ν is the viscosity, $\mathbf{f}(\mathbf{x})$ is the applied (divergence free) body force and the pressure p is determined by the condition of incompressibility,

$$\nabla \cdot \mathbf{u} = 0. \tag{9.2.2}$$

The vorticity is a scalar in $2d$

$$\omega = \hat{\mathbf{k}} \cdot \nabla \times \mathbf{u} = \frac{\partial u_2}{\partial x} - \frac{\partial u_1}{\partial y}, \tag{9.2.3}$$

and its evolution equation is

$$\frac{\partial \omega}{\partial t} + \mathbf{u} \cdot \nabla \omega = \nu \Delta \omega + \operatorname{curl} \mathbf{f}. \tag{9.2.4}$$

We consider the vorticity evolution (9.2.4) to be the defining equation, and invoke periodic boundary conditions on the square domain $\Omega \equiv [0, L]^2$. The velocity vector field is expressed in terms of the vorticity as

$$\mathbf{u} = \left(-\frac{\partial \Delta^{-1} \omega}{\partial y}, \frac{\partial \Delta^{-1} \omega}{\partial x} \right), \tag{9.2.5}$$

where, without loss of generality, we may assume that the spatially averaged vorticity vanishes so that the Laplacian may be appropriately inverted. Upper bounds on the attractor dimension for the system are determined by considering the time evolution of volume elements in the system's configuration space which, in this case, is $L^2(\Omega)$. If, say, all the N-dimensional volumes contract to zero volume as $t \to \infty$, then the attractor cannot contain any N-dimensional subsets and hence $d_L \leq N$. The goal is to determine the smallest possible N with this property, as it constitutes an upper bound on the dimension of the global attractor in both the fractal and Hausdorff sense.

We restrict our consideration to infinitesimal volume elements whose evolutions are controlled by the linearized Navier-Stokes equations, linearized about an arbitrary solution on the attractor. For (9.2.4), the linearized equation for the difference $\delta\omega$ between two neighboring solutions is

$$\frac{\partial \delta\omega}{\partial t} = -\mathbf{A}(t)\delta\omega = -\mathbf{u}\cdot\nabla\delta\omega - \delta\mathbf{u}\cdot\nabla\omega + \nu\Delta\delta\omega, \tag{9.2.6}$$

where the associated variation in the velocity is

$$\delta\mathbf{u} = \left(-\frac{\partial\Delta^{-1}\delta\omega}{\partial y}, \frac{\partial\Delta^{-1}\delta\omega}{\partial x}\right). \tag{9.2.7}$$

In $L^2(\Omega)$, these $\delta\omega_i$ form an N-dimensional volume element or parallelpiped of volume

$$V_N(t) = |\delta\omega_1(t)\wedge\ldots\wedge\delta\omega_N(t)|, \tag{9.2.8}$$

each edge of which develops according to (9.2.6). V_N itself evolves according to

$$V_N(t) = V_N(0)\exp\left(-\int_0^t Tr[\mathbf{A}(t')\mathbf{P}_N(t')]dt'\right). \tag{9.2.9}$$

In the above, $\mathbf{P}_N(t)$ is the orthogonal projection onto the *finite* dimensional linear subspace $\mathbf{P}_N L^2(\Omega)$. In order for N to be an upper bound on the attractor dimension, the volume elements $V_N(t)$ about *any* solution $\omega(t)$ on the attractor must vanish as $t\to\infty$. Rewriting (9.2.9),

$$\begin{aligned} V_N(t) &= V_N(0)\exp\left[-t\left(\frac{1}{t}\int_0^t Tr[\mathbf{A}(t')\mathbf{P}_N(t')]dt'\right)\right] \\ &\underset{t\to\infty}{\longrightarrow} V_N(0)\exp\{-t\,\langle Tr[\mathbf{AP}_N]\rangle\}, \end{aligned} \tag{9.2.10}$$

where $\langle\cdot\rangle$ denotes the largest possible time average. Thus, to determine an upper bound on the attractor dimension we look for the smallest[1] N which satisfies

$$\langle -Tr[\mathbf{AP}_N]\rangle < 0 \tag{9.2.11}$$

for all solutions $\omega(t)$. Let $\phi_1(t),\ldots,\phi_N(t)$ be orthonormal functions spanning $\mathbf{P}_N L^2(\Omega)$, and let the associated vector fields $\mathbf{v}_1(t),\ldots,\mathbf{v}_N(t)$ be

$$\mathbf{v}_n = \left(-\frac{\partial\Delta^{-1}\phi_n}{\partial y}, \frac{\partial\Delta^{-1}\phi_n}{\partial x}\right). \tag{9.2.12}$$

[1] In terms of the time average $\langle\cdot\rangle$, the sum of the first N global Lyapunov exponents (see (4.2.19) and (4.2.20) of Chapter 4) is related to $\langle Tr[\mathbf{AP}_N]\rangle$ by

$$\sum_{n=1}^{N}\mu_n = -\langle Tr[\mathbf{AP}_N]\rangle.$$

Estimate the trace as follows:

$$
\begin{aligned}
Tr[\mathbf{A}(t)\mathbf{P}_N(t)] &= \sum_{n=1}^{N}\int_\Omega \phi_n(t)\mathbf{A}(t)\phi_n(t)\,d^2x \\
&= \nu\sum_{n=1}^{N}\int_\Omega |\nabla\phi_n|^2\,d^2x + \sum_{n=1}^{N}\int_\Omega \phi_n(\mathbf{u}\cdot\nabla\phi_n + \mathbf{v}_n\cdot\nabla\omega)\,d^2x \\
&= \nu Tr[-\Delta\mathbf{P}_N] + \sum_{n=1}^{N}\int_\Omega \phi_n\mathbf{v}_n\cdot\nabla\omega\,d^2x, \qquad (9.2.13)
\end{aligned}
$$

where we have used the fact that $\mathbf{u}$ is divergence free to eliminate one of the terms. Because we know the spectrum of the Laplacian explicitly, the real work consists of finding good sharp upper bounds on the last sum in (9.2.13).

Using the Schwarz inequality, we have

$$
\left|\sum_{n=1}^{N}\int_\Omega \phi_n\mathbf{v}_n\cdot\nabla\omega\,d^2x\right| \le \int_\Omega\left(\sum_{n=1}^{N}\phi_n^2\right)^{1/2}\left(\sum_{n=1}^{N}|\mathbf{v}_n|^2\right)^{1/2}|\nabla\omega|\,d^2x. \qquad (9.2.14)
$$

Using a Hölder inequality we pull out the sum of the squares of the $\mathbf{v}_n$ in the L^∞ norm to obtain

$$
\left|\sum_{n=1}^{N}\int_\Omega \phi_n\mathbf{v}_n\cdot\nabla\omega\,d^2x\right| \le \left\|\sum_{n=1}^{N}|\mathbf{v}_n|^2\right\|_\infty^{1/2}\int_\Omega\left(\sum_{n=1}^{N}\phi_n^2\right)^{1/2}|\nabla\omega|\,d^2x. \qquad (9.2.15)
$$

The Cauchy-Schwarz inequality is now used to separate the two factors inside the integral above:

$$
\left|\sum_{n=1}^{N}\int_\Omega \phi_n\mathbf{v}_n\cdot\nabla\omega\,d^2x\right| \le \left\|\sum_{n=1}^{N}|\mathbf{v}_n|^2\right\|_\infty^{1/2}\left(\int_\Omega\sum_{n=1}^{N}\phi_n^2\,d^2x\right)^{1/2}\|\nabla\omega\|_2. \qquad (9.2.16)
$$

To estimate the first factor above we now use Constantin's theorem which provides L^∞ estimates on collections of functions whose *gradients* are orthonormal. We use it in the form where it is applicable to the sum of the squares of the $\mathbf{v}_n$'s.

Theorem 9.1 *If, for* $1 \le n \le N$, $\mathbf{v}_n$ *are functions whose gradients are orthonormal, that is* $\int_\Omega(\nabla v_{n\beta})\cdot(\nabla v_{m\beta})\,d^2x = \delta_{mn}$*; then*

$$
\left\|\sum_{n=1}^{N}|\mathbf{v}_n|^2\right\|_\infty \le c\left\{1+\log(L^2 Tr[-\Delta\mathbf{P}_N])\right\}, \qquad (9.2.17)
$$

where

$$Tr[-\Delta \mathbf{P}_N] = \sum_{n=1}^{N} \int_\Omega |\nabla \phi_n|^2 \, d^2x, \tag{9.2.18}$$

and the constant c is independent of N.

Proof: Consider vector valued functions $\mathbf{A}_\alpha(\mathbf{x})$ which are arbitrary linear combinations of the $\mathbf{v}_n(\mathbf{x})$ defined above

$$A_\alpha = \sum_{n=1}^{N} \xi_n v_{n\alpha} \tag{9.2.19}$$

with

$$v_{n\alpha} = -\varepsilon_{\alpha\beta} \partial_\beta \Delta^{-1} \phi_n. \tag{9.2.20}$$

Note that because the ϕ_n are orthonormal, the $\mathbf{v}_n$ have orthonormal gradients.

$$\int_\Omega (\nabla v_{n\beta}) \cdot (\nabla v_{m\beta}) \, d^2x = \delta_{nm}. \tag{9.2.21}$$

Then simply

$$\|\nabla \mathbf{A}\|_2^2 = \sum_{n=1}^{N} \xi_n^2 = |\xi|^2. \tag{9.2.22}$$

Now we apply Lemma 7.1 of Chapter 7 to $\mathbf{A}$:

$$\begin{aligned} \left| \sum_{n=1}^{N} \xi_n \mathbf{v}_{n\alpha} \right| &\le c\,|\xi| \left[1 + \log \left(L |\xi|^{-1} \left\| \sum_{n=1}^{N} \xi_n \Delta \mathbf{v}_n \right\|_2 \right) \right]^{1/2} \\ &\le c\,|\xi| \left[1 + \log \left(L^2 \sum_{n=1}^{N} \|\Delta \mathbf{v}_n\|_2^2 \right) \right]^{1/2}, \end{aligned} \tag{9.2.23}$$

where we have used the Schwarz inequality in the last step. We rewrite this as

$$|\xi|^{-1} \left| \sum_{n=1}^{N} \xi_n \mathbf{v}_n \right| \le c \left[1 + \log \left(L^2 \sum_{n=1}^{N} \|\Delta \mathbf{v}_n\|_2^2 \right) \right]^{1/2} \tag{9.2.24}$$

and because this is true for every N-vector ξ, we have

$$\left\| \sum_{n=1}^{N} |\mathbf{v}_n|^2 \right\|_\infty \le c \left[1 + \log \left(L^2 \sum_{n=1}^{N} \|\Delta \mathbf{v}_n\|_2^2 \right) \right]. \tag{9.2.25}$$

Using (9.2.20), we note finally that the sum of $\|\Delta \mathbf{v}_n\|_2^2$ inside the log term above is simply estimated,

$$\begin{aligned} \sum_{n=1}^{N} \|\Delta \mathbf{v}_n\|_2^2 &= \sum_{n=1}^{N} \int_\Omega |\Delta v_{n\beta}|^2 \, d^2x \\ &= \sum_{n=1}^{N} \int_\Omega (\varepsilon_{\beta\gamma}\partial_\gamma \phi_n)(\varepsilon_{\beta\delta}\partial_\delta \phi_n) \, d^2x \\ &= \sum_{n=1}^{N} \int_\Omega |\nabla \phi_n|^2 \, d^2x \\ &= Tr[-\Delta \mathbf{P}_N]. \end{aligned} \tag{9.2.26}$$

Inserting (9.2.26) above into (9.2.25) finishes the proof. □

Continuing with our estimation of (9.2.16) above, its middle factor is evaluated by recalling that the ϕ_n are orthonormal functions so that

$$\sum_{n=1}^{N} \int_\Omega \phi_n^2 \, d^2x = N. \tag{9.2.27}$$

The first term in (9.2.13), the trace of the Laplacian in an N-dimensional subspace, is easily estimated (see exercises 1 and 2)

$$Tr[-\Delta \mathbf{P}_N] \geq c\,L^{-2} N^{\frac{2+d}{d}}. \tag{9.2.28}$$

Hence we rewrite (9.2.27) as

$$\int_\Omega \sum_{n=1}^{N} \phi_n^2 \, d^2x \leq c\,(Tr[-\Delta \mathbf{P}_N])^{1/2}\, L. \tag{9.2.29}$$

Putting together equations (9.2.16), (9.2.17) and (9.2.29) we arrive at

$$\left| \sum_{n=1}^{N} \int_\Omega \phi_n \mathbf{v}_n \cdot \nabla\omega \, d^2x \right| \leq c \|\nabla\omega\|_2 \left[g\left(L^2 Tr[-\Delta \mathbf{P}_N] \right) \right]^{1/2}, \tag{9.2.30}$$

where $g(\zeta) = \sqrt{\zeta}(1 + \log \zeta)$. Taking the time average we have

$$\left\langle \left| \sum_{n-1}^{N} \int_\Omega \phi_n \mathbf{v}_n \cdot \nabla\omega \, d^2x \right| \right\rangle \leq c \left\langle \|\nabla\omega\|_2 \left[g\left(L^2 Tr[-\Delta \mathbf{P}_N] \right) \right]^{1/2} \right\rangle. \tag{9.2.31}$$

Using the Cauchy-Schwarz inequality on the time average on the right-hand side above gives

$$\left\langle\left|\sum_{n=1}^{N}\int_{\Omega}\phi_n\mathbf{v}_n\cdot\nabla\omega\,d^2x\right|\right\rangle\le c\left\langle\|\nabla\omega\|_2^2\right\rangle^{1/2}\left\langle g\left(L^2Tr[-\Delta\mathbf{P}_N]\right)\right\rangle^{1/2}. \tag{9.2.32}$$

Now, the function $g(\zeta)$ is concave for $\zeta > 1/e$. For large values of N which are appropriate for turbulent flows, $L^2Tr[-\Delta\mathbf{P}_N] \gg 1$, we invoke Jensen's inequality ($\langle g(\zeta)\rangle \le g(\langle\zeta\rangle)$ for g concave) to find

$$\left\langle\left|\sum_{n=1}^{N}\int_{\Omega}\phi_n\mathbf{v}_n\cdot\nabla\omega\,d^2x\right|\right\rangle\le c\left\langle\|\nabla\omega\|_2^2\right\rangle^{1/2}\left[g\left(\langle L^2Tr[-\Delta\mathbf{P}_N]\rangle\right)\right]^{1/2}. \tag{9.2.33}$$

Therefore, the time averaged trace in (9.2.10) and (9.2.11), controlling the exponential growth or contraction of volume elements, is estimated by

$$\begin{aligned}\langle Tr[\mathbf{AP}_N]\rangle \ge\ & \frac{\nu}{L^2}\left\langle L^2Tr\,[-\Delta\mathbf{P}_N]\right\rangle \\ & -c\left\langle\|\nabla\omega\|_2^2\right\rangle^{1/2}\left[g\left(\langle L^2Tr\,[-\Delta\mathbf{P}_N]\rangle\right)\right]^{1/2}.\end{aligned} \tag{9.2.34}$$

Next we use the $2d$ Grashof number $\mathcal{G}$, defined in Chapter 2 and Table 6.2:

$$\mathcal{G} = L^2\|\mathbf{f}\|_2/\nu^2. \tag{9.2.35}$$

Multiplying the vorticity version of the Navier-Stokes equation (9.2.4) by ω, integrating over the spatial variables and taking the time average, we have

$$\nu\left\langle\|\nabla\omega\|_2^2\right\rangle = \left\langle\int_{\Omega}\omega\hat{\mathbf{k}}\cdot\operatorname{curl}\mathbf{f}\,d^2x\right\rangle. \tag{9.2.36}$$

A simple integration by parts and application of Cauchy's inequality yields

$$\nu\left\langle\|\nabla\omega\|_2^2\right\rangle \le \langle\|\nabla\omega\|_2\rangle\,\|\mathbf{f}\|_2 \le \left\langle\|\nabla\omega\|_2^2\right\rangle^{1/2}\|\mathbf{f}\|_2, \tag{9.2.37}$$

so that

$$\left\langle\|\nabla\omega\|_2^2\right\rangle^{1/2} \le \|\mathbf{f}\|_2/\nu = \nu\mathcal{G}/L^2. \tag{9.2.38}$$

Define $\tilde{N}$ by

$$\tilde{N}^2 = L^2\langle Tr[-\Delta\mathbf{P}_N]\rangle. \tag{9.2.39}$$

Then the trace formula in (9.2.34) is

$$\langle Tr[\mathbf{AP}_N]\rangle \ge \frac{\nu}{L^2}\left[\tilde{N}^2 - c\,\mathcal{G}\,\tilde{N}^{1/2}(1+\log\tilde{N})^{1/2}\right]. \tag{9.2.40}$$

As N increases, then so does $\tilde{N}$ and for a given $\mathscr{G}$ it eventually forces the right-hand side of (9.2.40) to become positive so that all volume elements of dimension higher than N contract to zero. If the logarithmic term were absent in (9.2.40), then this crossover point would occur when $\tilde{N} \sim \mathscr{G}^{2/3}$. The logarithm introduces corrections to this. The precise answer, including the logarithmic correction, leads us to conclude that all N-dimensional volume elements contract to zero when

$$N \geq c\,\mathscr{G}^{2/3}(1+\log\mathscr{G})^{1/3}. \tag{9.2.41}$$

To establish this result, we must show that

$$\tilde{N} \geq c\,\mathscr{G}^{2/3}(1+\log\mathscr{G})^{1/3} \quad \Rightarrow \quad \tilde{N}^2 - c'\,\mathscr{G}\tilde{N}^{1/2}(1+\log\tilde{N})^{1/2} \geq 0. \tag{9.2.42}$$

We prove this via transposition by showing that

$$\tilde{N}^2 - c'\,\mathscr{G}\tilde{N}^{1/2}(1+\log\tilde{N})^{1/2} < 0 \quad \Rightarrow \quad \tilde{N} < c\,\mathscr{G}^{2/3}(1+\log\mathscr{G})^{1/3}. \tag{9.2.43}$$

Toward this end, assume that

$$\tilde{N}^2 - c'\,\mathscr{G}\tilde{N}^{1/2}(1+\log\tilde{N})^{1/2} < 0. \tag{9.2.44}$$

Then

$$\frac{\tilde{N}^3}{1+\log\tilde{N}} < c''\mathscr{G}^2, \tag{9.2.45}$$

and

$$3\log\tilde{N} - \log(1+\log\tilde{N}) < 2\log\mathscr{G} + \log c''. \tag{9.2.46}$$

But it is elementary to show that for the relevant regime where $\tilde{N} > 1$,

$$\log(1+\log\tilde{N}) \leq \log\tilde{N} \tag{9.2.47}$$

so (9.2.46) implies

$$1+\log\tilde{N} < \tilde{c}(1+\log\mathscr{G}). \tag{9.2.48}$$

Now, the premise (9.2.44) is

$$\tilde{N}^{3/2} < c'\mathscr{G}(1+\log\tilde{N})^{1/2}, \tag{9.2.49}$$

and inserting (9.2.48) we have

$$\tilde{N}^{3/2} < c'\sqrt{\tilde{c}}\,\mathscr{G}(1+\log\mathscr{G})^{1/2}. \tag{9.2.50}$$

This is equivalent to the right-hand side of (9.2.43) and establishes the result.

Thus the attractor dimension is bounded by

$$d_L \leq c\,\mathscr{G}^{2/3}(1+\log\mathscr{G})^{1/3}. \tag{9.2.51}$$

Finally, to express d_L in terms of the Kraichnan length λ_{K_r} introduced in Chapter 3, instead of $\mathscr{G}$, we return to (9.2.34) and express $\langle \|\nabla\omega\|_2^2 \rangle$ in terms of the average enstrophy dissipation rate χ given in (3.3.32): $\chi = \nu L^{-2} \langle \|\nabla\omega\|_2^2 \rangle$. From (3.3.34), the definition of λ_{K_r} is given by $\lambda_{K_r} = \left(\nu^3/\chi\right)^{1/6}$, so we finally obtain

$$d_L \leq c \left(\frac{L}{\lambda_{K_r}}\right)^2 \left[1 + \log\left(\frac{L}{\lambda_{K_r}}\right)\right]^{1/3}. \tag{9.2.52}$$

Apart from the logarithmic correction, the estimate in (9.2.52) fits with the scaling theory of $2d$ turbulence, that is, the number of degrees of freedom in a $2d$ domain is $\left(L/\lambda_{K_r}\right)^2$. Note also that the estimate (9.2.51) is the same as (7.2.32) in section 7.2 for the first of the nontrivial number of degrees of freedom $\mathcal{N}_{3,1}$. This result shows that the rigorous methods of attractor dimension analysis, using the full properties of the $2d$ Navier-Stokes equations without uncontrolled approximations, coincide with the predictions arising from heuristic considerations.

9.3 The $3d$ attractor dimension estimate

The reader who has spent time in Chapter 7 will have understood that the lack of a regularity proof for the $3d$ Navier-Stokes equations precludes us from asserting the existence of a global attractor. This does not prevent us using the the Kolmogorov length λ_K, though, as it is a properly bounded quantity, being dependent on $\langle F_1 \rangle$. We recall that the energy dissipation rate and the Kolmogorov length are respectively defined by $\varepsilon = \nu L^{-3} \langle H_1 \rangle$ and $\lambda_K^{-1} = \left(\varepsilon/\nu^3\right)^{1/4}$. Because of the lack of a solid proof of the existence of a global attractor, the device used in Chapter 7 to circumvent this problem (see (7.4.4)) was to introduce $\sup_t H_1$ instead of the time average $\langle H_1 \rangle$ (or $\langle F_1 \rangle$) in order to define an alternative energy dissipation rate and an alternative Kolmogorov length according to

$$\bar{\varepsilon} = \nu L^{-3} \left(\sup_t H_1\right), \qquad \text{and} \qquad \Lambda_K^{-1} = \left(\frac{\bar{\varepsilon}}{\nu^3}\right)^{1/4}. \tag{9.3.1}$$

We assume, therefore, that $\sup_t H_1$ is bounded and proceed to estimate the attractor dimension based on Λ_K. Turning to the trace formula, no advantage is gained in $3d$ by working with $\omega \in L^2$. Hence we consider the velocity field orbit $\mathbf{u}(t) \in L^2$. The linearized evolution of $\delta\mathbf{u}$ about $\mathbf{u}$ is

$$\delta\mathbf{u}_t + \mathbf{u} \cdot \nabla\delta\mathbf{u} + \delta\mathbf{u} \cdot \nabla\mathbf{u} = \nu\Delta\delta\mathbf{u} - \nabla\delta p \tag{9.3.2}$$

which can formally be written $\delta\mathbf{u}_t = -\mathbf{A}\delta\mathbf{u}$. As in (9.2.11), we want to find the value of N that turns the sign of $\langle Tr\,[\mathbf{AP}_N]\rangle$ from negative to positive. This value of N, ($\tilde{N}$ say) bounds above d_L. The orthonormal basis $\{\boldsymbol{\phi}_1, \boldsymbol{\phi}_2, \ldots, \boldsymbol{\phi}_N\}$ in L^2 consists of divergence-free vector functions. The trace formula we need to explicitly estimate is

$$Tr\,[\mathbf{AP}_N] = -\sum_{n=1}^{N}\int_\Omega \boldsymbol{\phi}_n\cdot\{\nu\Delta\boldsymbol{\phi}_n - \mathbf{u}\cdot\nabla\boldsymbol{\phi}_n - \boldsymbol{\phi}_n\cdot\nabla\mathbf{u} - \nabla\tilde{p}(\boldsymbol{\phi}_n)\}\,dV. \tag{9.3.3}$$

Since $\text{div}\boldsymbol{\phi}_n = 0$, the pressure term integrates away, as does the second term, to give

$$Tr\,[\mathbf{AP}_N] \geq \nu\sum_{n=1}^{N}\int_\Omega |\nabla\boldsymbol{\phi}_n|^2\,dV - \sum_{n=1}^{N}\int_\Omega |\nabla u|\,|\boldsymbol{\phi}_n|^2\,dV. \tag{9.3.4}$$

The $\boldsymbol{\phi}_n$ satisfy what are known as Lieb-Thirring inequalities for orthonormal functions (see the Appendix)

$$\int_\Omega\left(\sum_{n=1}^{N}|\boldsymbol{\phi}_n|^2\right)^{\frac{5}{3}} d^3x \leq c\sum_{n=1}^{N}\int_\Omega |\nabla\boldsymbol{\phi}_n|^2\,d^3x, \tag{9.3.5}$$

where c is independent of N. Moreover, from (9.2.28), we know that $Tr\,[(-\Delta)\mathbf{P}_N]$ is bounded below (see also exercises 1 and 2). We can now exploit the Lieb-Thirring inequality (9.3.5) and estimate the last term in (9.3.4) as

$$\sum_{n=1}^{N}\int_\Omega |\mathbf{Du}|\,|\boldsymbol{\phi}_n|^2\,dV \leq \left[\int_\Omega |\mathbf{Du}|^{5/2}\,dV\right]^{2/5}\left[\int_\Omega\left(\sum_{n=1}^{N}|\boldsymbol{\phi}_n|^2\right)^{5/3} dV\right]^{3/5}. \tag{9.3.6}$$

Hence, using (9.3.5) and time averaging $\langle\cdot\rangle$, we find that (9.3.4) can be written as

$$\langle Tr\,[\mathbf{AP}_N]\rangle \geq \frac{2}{5}\nu\,\langle Tr\,[(-\Delta)\mathbf{P}_N]\rangle - c\,\nu^{-3/2}\left\langle\int_\Omega |\mathbf{Du}|^{5/2}\,dV\right\rangle. \tag{9.3.7}$$

The problem lies in the $\left\langle\|\mathbf{Du}\|_{5/2}^{5/2}\right\rangle$ term for which we have no a priori estimate, so we rewrite (9.3.7) as

$$\langle Tr\,[\mathbf{AP}_N]\rangle \geq \frac{2}{5}\nu\,\langle Tr\,[(-\Delta)\mathbf{P}_N]\rangle - c\,\nu^{-3/2}\left\langle\|\mathbf{Du}\|_\infty^{1/2}\right\rangle\left(\sup_t H_1\right). \tag{9.3.8}$$

We have an estimate for $\left\langle \|D\mathbf{u}\|_\infty^{1/2} \right\rangle$ from (7.3.32) so (9.3.8) becomes

$$\langle Tr\,[\mathbf{AP}_N]\rangle \geq c_1 \nu L^{-2} N^{5/3} - c_2 \nu L^2 \left(\frac{L}{\lambda_K}\right)^4 \Lambda_K^{-4} \tag{9.3.9}$$

where we have used the definition of Λ_K given in (9.3.1). Hence, the trace becomes positive when $N \geq \tilde{N}$ and so

$$d_L(\mathscr{A}) \leq \tilde{N} = c \left(\frac{L}{\Lambda_K}\right)^{12/5} \left(\frac{L}{\lambda_K}\right)^{12/5}. \tag{9.3.10}$$

If d_L is also associated with the number of degrees of freedom $\mathscr{N}$ of the system discussed in section 7.1, then this can be written in terms of the natural small scale of the system ℓ

$$\frac{L}{\ell} \leq c \left(\frac{L}{\Lambda_K}\right)^{1.6}. \tag{9.3.11}$$

where we have exploited the fact that $\lambda_K^{-1} \leq \Lambda_K^{-1}$. The exponent 1.6 is significantly closer to the conventional heuristic estimate of unity and it is certainly much better than the exponent 4 found in (7.4.11) and in (8.4.16).

9.4 References and further reading

The computation of volume elements and how they are related to the various dimensions (Lyapunov, Hausdorff, and capacity) can be found in the paper by Constantin and Foias [11] (see also Constantin, Foias, and Temam [43]) and are also discussed extensively in the books by Constantin and Foias [12] and Temam [14]. The $2d$ dimension estimate $\mathscr{G}^{2/3}(1+\log\mathscr{G})^{1/3}$ was first announced in Constantin, Foias, and Temam [44]. Moreover, Babin and Vishik [13] give arguments for a lower bound on this dimension which is proportional to $\mathscr{G}^{2/3}$. Constantin's theorem (Theorem 9.1) can be found in [45]. The calculation in section 9.2 follows that of Doering and Gibbon [46] who reworked that of Constantin, Foias, and Temam [44] using the vorticity instead of the velocity field formulation. In section 9.3 the exponent of 1.6 for the $3d$ case can be found in Gibbon and Titi [47] improving the original estimates in [43]. The related concept of determining modes is explained in Constantin, Foias, Manley and Temam [18, 43] — estimates and references can be found in Jones and Titi [48]. See Ilyin [49] for attractor dimension estimates on $2d$ manifolds.

Exercises

1 Show that

$$Tr[-\Delta \mathbf{P}_N] \geq c\,L^{-2}N^{\frac{2+d}{d}}, \tag{E9.1}$$

by considering the eigenvalues of the Stokes' operator $-\Delta$.

2 Use the inequality of Lieb and Thirring given in (9.3.5) to also prove (E9.1).

3 Consider the PDE known as the complex Ginzburg Landau (CGL) equation

$$u_t = Ru + (1+i\nu)\Delta u - (1+i\mu)u|u|^2 \tag{E9.2}$$

on the $2d$ periodic domain $\Omega \equiv [0,1]^2$. The parameters R, μ and ν are real and lie in the range $R > 1$, $|\mu| < \infty$ and $|\nu| < \infty$. Show that (i) when $|\mu| \leq \sqrt{3}$, then $d_L \leq R$ whereas (ii) when $|\mu| > \sqrt{3}$ then $d_L \leq c\,R$ where c is independent of ν but not μ.

10

Energy dissipation rate estimates for boundary-driven flows

10.1 Introduction

The four preceding chapters concentrated on the analysis of solutions of the Navier-Stokes equations in the absence of boundaries. The issue of regularity was the central concern there, and it was shown that a crucial quantity is $\|\nabla\mathbf{u}\|_2^2$, the sum of the squares of the integrals of the derivatives of the velocity field. The construction of weak solutions outlined in Chapter 5 provides explicit upper bounds on the time integral of $\|\nabla\mathbf{u}\|_2^2$, although no assurance was found that this remains finite at every instant of time. The small length scales in the solutions may be estimated in terms of this quantity (Chapters 7 and 8) and the system's global attractor – if it exists – is controlled also by this quantity (Chapter 9). In those cases it is not just the average value of $\|\nabla\mathbf{u}\|_2^2$ that is important: As long as this quantity remains finite pointwise in time for solutions of the Navier-Stokes equations, then the solutions are actually infinitely differentiable. In $2d$ we are able to produce a priori bounds on this quantity at each instant of time, but in $3d$ the mathematical manifestations of the vortex stretching mechanism prevent the same conclusion from being drawn.

This same quantity is proportional to the instantaneous rate of viscous energy dissipation in the flow. As introduced in Chapter 3, the time averaged rate of viscous energy dissipation per unit mass is

$$\varepsilon = \Omega^{-1}\nu\left\langle\|\nabla\mathbf{u}\|_2^2\right\rangle, \tag{10.1.1}$$

where Ω is the system volume and $\langle\cdot\rangle$ denotes a time average. This is of great interest for applications: In a nonequilibrium steady state the rate of energy dissipation must be balanced by the rate at which external work is done on the system and this, in turn, is directly related to the magnitude of the external forces or constraints that must be applied to the fluid in order to maintain the steady state. When these forces can be identified as physical drag forces, knowledge of the energy dissipation rate leads to estimates of the drag forces exerted by the fluid. More generally, the rate of energy dissipation is often related to the global transport or flux of some physical quantity. In the case of drag forces it is momentum being transported between boundaries, while for pipe or channel flow problems, it is mass being transported by the flow. In thermal convection problems it is heat which is transported by the flow and the heat flux is related to the energy dissipation rate. If we are content to consider time averaged quantities, then in some cases we may make some rigorous quantitative statements about the relationships between the magnitudes of these forces and fluxes.

In this chapter we will concentrate on obtaining explicit bounds on ε for two specific fundamental flow configurations. The approach developed here is to formulate variational principles for bounds on ε – both lower and upper bounds valid for laminar and turbulent flows – directly from the equations of motion without any additional assumptions, closures, or hypotheses. In section 10.2 we consider boundary-driven shear flow, which is studied in Chapter 3 from the approximate viewpoint of the conventional statistical theory of turbulence. Here we present a rigorous analysis of the problem. Thermal convection is the topic of section 10.3, and the analysis provides a rigorous relationship between the Rayleigh number (the nondimensional measure of the applied temperature gradient) and the Nusselt number (the nondimensional measure of the convective heat transfer). In section 10.4 we make some general observations and remarks on the method developed for these problems, and we provide comparisons of the results of this rigorous analysis with approximate theory and related experiments.

10.2 Boundary-driven shear flow

Consider the problem of an incompressible Newtonian fluid contained between rigid parallel plates located at $z = 0$ and $z = h$, as illustrated in Figure 3.1. The x-coordinate lies between 0 and L_1, the y-coordinate lies between 0 and L_2, and we impose periodic boundary conditions in the x- and y-directions. The fluid is driven by the boundary at $z = h$ moving in the x-direction at speed U, so the no-slip boundary conditions at the plates are $\mathbf{u}(x, y, 0, t) = 0$ and $\mathbf{u}(x, y, h, t) = \hat{\imath}U$. The Reynolds number is defined as $R = Uh/\nu$.

This is the same problem that we studied in section 3.2 in terms of statistical turbulence theory. We will focus on the 3*d* case here but the same problem can be formulated and analyzed in 2*d* by restricting **u** to the $x-z$ plane and removing the y dependence from all quantities. In the following we will manipulate the equations of motion as if they admitted strong solutions, reserving considerations of existence and regularity for section 10.4.

The analysis is based on the incompressible Navier-Stokes equations

$$\frac{\partial \mathbf{u}}{\partial t} + \mathbf{u} \cdot \nabla \mathbf{u} + \nabla p = \nu \Delta \mathbf{u}, \tag{10.2.1}$$

$$\nabla \cdot \mathbf{u} = 0. \tag{10.2.2}$$

We have chosen mass units so that the density $\rho = 1$. Taking the scalar product of **u** with the Navier-Stokes equations and integrating over the domain, after an appropriate integration by parts, we find that the energy equation becomes

$$\frac{1}{2}\frac{d}{dt}\|\mathbf{u}\|_2^2 = -\nu\|\nabla \mathbf{u}\|_2^2 + \nu U \int_0^{L_1} dx \int_0^{L_2} dy \left.\frac{\partial u_1}{\partial z}\right|_{z=h}. \tag{10.2.3}$$

This energy evolution equation asserts that the instantaneous rate of change of the kinetic energy in the fluid is equal to the loss of energy from viscous dissipation plus the rate at which work is performed against the viscous drag by the agent imposing the boundary condition. The goal is to obtain explicit upper bounds on the average energy dissipation rate

$$\varepsilon = \frac{\nu \left\langle \|\nabla \mathbf{u}\|_2^2 \right\rangle}{L_1 L_2 h}, \tag{10.2.4}$$

in terms of the system parameters (U, ν, h, L_1, L_2), directly from the Navier-Stokes equations. The time average $\langle \cdot \rangle$ defined in (6.3.10) is the largest possible time average for any initial condition.

In a steady state, the time average of (10.2.3) yields the balance

$$\nu \left\langle \|\nabla \mathbf{u}\|_2^2 \right\rangle = \nu U \int_0^{L_1} dx \int_0^{L_2} dy \left.\left\langle \frac{\partial u_1}{\partial z} \right\rangle\right|_{z=h} \tag{10.2.5}$$

between the largest possible energy dissipation rate and the product of the plate velocity U and the largest possible drag force exerted by the fluid on the plate. This identity, however, does not allow us to estimate ε solely in terms of the system parameters because it is not a closed relationship for ε. The problem is that we have not a priori connected the $\|\nabla \mathbf{u}\|_2^2$ with the shear rate on the boundary.

A lower bound on the energy dissipation rate can easily be formulated as a variational problem. Because any solution $\mathbf{u}(\mathbf{x},t)$ is a divergence-free vector field satisfying the boundary conditions, we have

$$\nu\|\nabla\mathbf{u}\|_2^2 \geq \inf\left\{\nu\|\nabla\mathbf{U}\|_2^2 \;\middle|\; \nabla\cdot\mathbf{U}=0, \quad \mathbf{U} \text{ satisfies the BCs.}\right\}. \quad (10.2.6)$$

This is the formulation of the obvious statement that the smallest possible value of the energy dissipation rate for a solution of the Navier-Stokes equations is at least as large as the smallest possible value amongst all divergence-free vector fields that satisfy the boundary conditions, whether or not they satisfy the dynamical equations.

The minimization problem in (10.2.6) may be attacked by standard techniques of the calculus of variations. We seek the minimum value of the functional

$$\mathbf{F}\{\mathbf{U}\} = \nu \int |\nabla\mathbf{U}|^2 \, d^3x, \quad (10.2.7)$$

subject to the constraint $\nabla\cdot\mathbf{U}=0$ at each point in space. The divergence-free constraint may be imposed by the method of Lagrange multipliers. Denoting the Lagrange multiplier by $-2p(\mathbf{x})$ and replacing $\mathbf{F}\{\mathbf{U}\}$ with the functional

$$\mathbf{F}\{\mathbf{U}\} = \int \left\{\nu\,|\nabla\mathbf{U}|^2 - 2p\nabla\cdot\mathbf{U}\right\} d^3x, \quad (10.2.8)$$

the associated Euler-Lagrange equations are

$$0 = \frac{\delta\mathbf{F}}{\delta\mathbf{U}} = -2\nu\Delta\mathbf{U} + 2\nabla p. \quad (10.2.9)$$

Supplemented with the divergence-free constraint, then, the minimum is realized by the solution of the *Stokes equations*

$$\nu\Delta\mathbf{U} = \nabla p, \quad (10.2.10)$$

$$\nabla\cdot\mathbf{U} = 0, \quad (10.2.11)$$

with periodic boundary conditions in the x- and y-directions, and the no-slip conditions, $\mathbf{U}(x,y,0)=0$ and $\mathbf{U}(x,y,h)=\hat{\mathbf{i}}U$, at the plates. The Stokes equations are the linearized, stationary Navier-Stokes equations, and their solution yields the smallest possible energy dissipation rate for the geometry.

The solution of the Stokes equations (10.2.10) and (10.2.11) is planar Couette flow

$$\mathbf{U} = \hat{\mathbf{i}}\frac{Uz}{h}, \quad (10.2.12)$$

which is additionally a stationary solution of the Navier-Stokes equations (because the $\mathbf{U} \cdot \nabla \mathbf{U}$ vanishes identically for this vector field and we may take $p =$ constant for this flow). The minimum energy dissipation rate is thus

$$\varepsilon_{min} = \frac{\nu U^2}{h^2}. \tag{10.2.13}$$

The force required to slide the top plate over the bottom lubricated by the fluid in this laminar state is (recall (3.2.18)) the total energy dissipation rate per unit velocity, which is also the wall shear stress τ times the area $A = L_1 L_2$ of the plates:

$$F_{min} = \frac{\varepsilon_{min} h A}{U} = \tau_{min} \times A = \frac{\nu U A}{h}. \tag{10.2.14}$$

As mentioned in Chapter 3, this is the familiar expression for the drag force from elementary physics texts, proportional to the shear rate, the contact area and the viscosity.

It is useful to express this force in nondimensional terms. Measuring lengths, masses, and time in units of the geometric parameter h and the material parameters ρ and ν, we have

$$\frac{h^2 \tau}{\rho \nu^2} = \frac{Uh}{\nu} = R. \tag{10.2.15}$$

(The density is included solely as a dimensional check.) The laminar Couette flow is nonlinearly stable at low Reynolds numbers, so the minimum dissipation rate and the corresponding minimum drag are realized at low R.

At higher R other time asymptotic states, including turbulent ones, may be possible, in which case the lower bound will no longer be realized. We cannot formulate a variational upper bound on ε simply by replacing the inf by a sup, because this would result in an infinite upper bound. Nor can we immediately proceed along the lines of previous chapters by using functional inequalities directly on the energy evolution equations (10.2.3) because we cannot necessarily find appropriate estimates for a function on a boundary of a region in terms of estimates for the inner bulk of that region.

These problems may be circumnavigated by employing a change of variables, which we will refer to as the *background flow* technique. In the spirit of the Reynolds decomposition into mean and fluctuating components, we decompose the velocity as

$$\mathbf{u}(x, y, z, t) = \hat{\mathbf{i}}\phi(z) + \mathbf{v}(x, y, z, t), \tag{10.2.16}$$

where the boundary conditions are contained in the *background flow* $\hat{\mathbf{i}}\phi(z)$, $\phi(0) = 0$ and $\phi(h) = U$, and the dynamics are contained in the *fluctuation field* $\mathbf{v}$. We make no claims with regard to any possible relationship between the background flow and some average of $\mathbf{u}$; indeed, there is no guarantee that $\mathbf{u}$ possesses a time average or that some averages exist that do not depend on initial data, etc.

The fluctuation field satisfies

$$\frac{\partial \mathbf{v}(\mathbf{x},t)}{\partial t} + \mathbf{v}\cdot\nabla\mathbf{v} + \hat{\mathbf{i}}\phi' v_3 + \phi\frac{\partial}{\partial x}\mathbf{v} + \nabla p = \nu\Delta\mathbf{v} + \hat{\mathbf{i}}\nu\phi'', \tag{10.2.17}$$

$$\nabla\cdot\mathbf{v} = 0, \tag{10.2.18}$$

where ϕ' and ϕ'' denote derivatives of $\phi(z)$. The boundary conditions for $\mathbf{v}$ are periodic in the x- and y-directions with periods L_1 and L_2 respectively, whereas in the z direction $\mathbf{v}(x,y,0,t) = 0 = \mathbf{v}(x,y,h,t)$. The background flow profile is essentially arbitrary at this stage, constrained only by the boundary conditions.

The energy equation for $\mathbf{v}$, obtained by taking the scalar product of $\mathbf{v}$ with (10.2.17), integrating, and then integrating by parts, is

$$\begin{aligned}\frac{1}{2}\frac{d}{dt}\|\mathbf{v}\|_2^2 &= -\nu\|\nabla\mathbf{v}\|_2^2 - \int_0^{L_1}dx\int_0^{L_2}dy\int_0^h dz\,\phi' v_1 v_3 \\ &\quad -\nu\int_0^{L_1}dx\int_0^{L_2}dy\int_0^h dz\,\phi'\frac{\partial v_1}{\partial z}.\end{aligned} \tag{10.2.19}$$

The energy dissipation in $\mathbf{u}$ is expressed in terms of $\mathbf{v}$ and ϕ as

$$\nu\|\nabla\mathbf{u}\|_2^2 = \nu\|\nabla\mathbf{v}\|_2^2 + 2\nu\int_0^{L_1}dx\int_0^{L_2}dy\int_0^h dz\,\phi'\frac{\partial v_1}{\partial z} + \nu\|\phi'\|_2^2. \tag{10.2.20}$$

Replacing half of the first term on the right-hand side of (10.2.19), together with the last term, by their counterparts in (10.2.20), the energy evolution equation for $\mathbf{v}$ becomes

$$\frac{1}{2}\frac{d}{dt}\|\mathbf{v}\|_2^2 = -\frac{1}{2}\nu\|\nabla\mathbf{v}\|_2^2 - \int_0^{L_1}dx\int_0^{L_2}dy\int_0^h dz\,\phi' v_1 v_3 - \frac{1}{2}\nu\|\nabla\mathbf{u}\|_2^2 + \frac{1}{2}\nu\|\phi'\|_2^2. \tag{10.2.21}$$

Rearranging the terms in (10.2.21) we find

$$\frac{d}{dt}\|\mathbf{v}\|_2^2 + \nu\|\nabla\mathbf{u}\|_2^2 = \nu\|\phi'\|_2^2 - 2\int_0^{L_1}dx\int_0^{L_2}dy\int_0^h dz\,\left\{\frac{\nu}{2}|\nabla\mathbf{v}|^2 + \phi' v_1 v_3\right\}. \tag{10.2.22}$$

Now, *if* the last integral on the right-hand side of (10.2.22) is always positive for our choice of ϕ, then we may neglect it by turning the differential equation into a differential inequality:

$$\frac{d}{dt}\|\mathbf{v}\|_2^2 + \nu\|\nabla\mathbf{u}\|_2^2 \leq \nu\|\phi'\|_2^2. \tag{10.2.23}$$

Then, averaging over time from 0 to T and remembering that ϕ does *not* depend on time, we have

$$\frac{1}{T}\|\mathbf{v}(\cdot,T)\|_2^2 + \frac{\nu}{T}\int_0^T \|\nabla\mathbf{u}(\cdot,t)\|_2^2\,dt \leq \nu\|\phi'\|_2^2 + \frac{1}{T}\|\mathbf{v}(\cdot,0)\|_2^2. \tag{10.2.24}$$

When the initial condition $\|\mathbf{v}(\cdot,0)\|_2^2$ is square integrable, as will be presumed, the last term on the right-hand side of (10.2.24) vanishes as $T \to \infty$. Moreover, because the first term on the left is manifestly positive it may be dropped without violating the inequality. We could then conclude that

$$\varepsilon = \frac{\nu\left\langle\|\nabla\mathbf{u}(\cdot,t)\|_2^2\right\rangle}{hA} \leq \frac{\nu\|\phi'\|_2^2}{hA} = \frac{\nu}{h}\int_0^h |\phi'(z)|^2\,dz. \tag{10.2.25}$$

We stress that (10.2.25) is valid under the conditions that $\phi(z)$ satisfies the boundary conditions $\phi(0) = 0$ and $\phi(h) = U$, *and* that

$$\int_0^{L_1} dx \int_0^{L_2} dy \int_0^h dz\,\left\{\frac{\nu}{2}|\nabla\mathbf{v}|^2 + \phi' v_1 v_3\right\} \geq 0 \tag{10.2.26}$$

for the solutions $\mathbf{v}(x,y,z,t)$ of the fluctuation's dynamical equation. This latter condition will certainly be the case if ϕ is chosen so that (10.2.26) holds for *all* divergence-free vector fields $\mathbf{v}(x,y,z)$ vanishing at $z = 0$ and $z = h$. The best (lowest) upper bound on ε obtained in this way will then come from an optimal choice of ϕ.

These considerations can be collected together in the form of a constrained variational problem:

$$\varepsilon \leq \inf\left\{\frac{\nu}{h}\int_0^h |\phi'(z)|^2\,dz \,\middle|\, \phi(0) = 0,\ \phi(h) = U \text{ and } G_\phi\{\mathbf{v}\} \geq 0\right\}, \tag{10.2.27}$$

where the functional $G_\phi\{\mathbf{v}\}$ is

$$G_\phi\{\mathbf{v}\} = \int_0^{L_1} dx \int_0^{L_2} dy \int_0^h dz\,\left\{\frac{\nu}{2}|\nabla\mathbf{v}|^2 + \phi' v_1 v_3\right\}, \tag{10.2.28}$$

so defined for divergence-free vector fields $\mathbf{v}(\mathbf{x})$ which are periodic in x and y, and vanishing at $z = 0$ and $z = h$. This constitutes a constrained variational problem for the optimal profile $\phi(z)$.

We refer to the constraint $G_\phi\{\mathbf{v}\} \geq 0$ as a *spectral constraint* because it is equivalently stated as the requirement that an associated self-adjoint eigenvalue problem has a positive spectrum. Indeed, the possible signs of $G_\phi\{\mathbf{v}\}$ are those of the homogeneous functional

$$K_\phi\{\mathbf{v}\} = \frac{G_\phi\{\mathbf{v}\}}{\frac{1}{2}\|\mathbf{v}\|_2^2}, \tag{10.2.29}$$

and the nonnegativity of the quadratic form in $G_\phi\{\mathbf{v}\}$ is equivalent to $K_\phi\{\mathbf{v}\}$ being bounded from below. The minimum value of $K_\phi\{\mathbf{v}\}$ is realized by the solution of the Euler-Lagrange equations

$$0 = \frac{\delta K_\phi\{\mathbf{v}\}}{\delta \mathbf{v}}, \tag{10.2.30}$$

which constitute the eigenvalue problem

$$\begin{aligned}
\lambda v_1 &= -\nu\Delta v_1 + \frac{\partial p}{\partial x} + \phi'(z)v_3 \\
\lambda v_2 &= -\nu\Delta v_2 + \frac{\partial p}{\partial y} \\
\lambda v_3 &= -\nu\Delta v_3 + \frac{\partial p}{\partial z} + \phi'(z)v_1 \\
0 &= \frac{\partial v_1}{\partial x} + \frac{\partial v_2}{\partial y} + \frac{\partial v_3}{\partial z},
\end{aligned} \tag{10.2.31}$$

for the eigenvalue λ and eigenvector $\mathbf{v}$, periodic in x and y and vanishing at $z = 0$ and $z = h$. As usual, the "pressure" p in (10.2.31) is the Lagrange multiplier for the divergence-free constraint.

When the smallest eigenvalue λ of the self-adjoint problem in (10.2.31) is nonnegative, then K_ϕ is bounded below and $G_\phi(\mathbf{v})$ is a nonnegative functional. This spectral problem is precisely the sort of eigenvalue equation encountered in the discussion of nonlinear stability in section 2.2. In that context, the condition $G_\phi(\mathbf{v}) > 0$ is equivalent to nonlinear stability of the flow field $\hat{\mathbf{i}}\phi(z)$ for a fluid of viscosity $\nu/2$. Note, however, that in contrast to the considerations of nonlinear stability of a stationary flow, the background field is not necessarily a solution of the Navier-Stokes equations.

Although the exact solution of the minimization problem for the upper bound on ε in (10.2.27) is a difficult problem, the utility of this formulation is apparent when we point out that $\nu\|\phi'\|_2^2/hA$ will be an upper bound on ε whenever a trial profile ϕ satisfies the boundary conditions and

$G_\phi\{\mathbf{v}\} \geq 0$. Of course, most acceptable "trial" profiles will not yield the best upper bound, but they will produce an upper bound nonetheless. The quality of the estimates derived in this way can be determined by other considerations, as will be discussed in section 10.4. The challenge right now is to manufacture appropriate trial background flows.

Consider the first trial choice $\phi(z) = Uz/h$, which is the Couette profile. Its energy dissipation rate is a lower bound, and it is also an upper bound if $G_\phi \geq 0$, i.e., if

$$\int_0^{L_1} dx \int_0^{L_2} dy \int_0^h dz \left\{ \frac{1}{2}\nu|\nabla\mathbf{v}|^2 + \frac{U}{h}v_1 v_3 \right\} \geq 0 \tag{10.2.32}$$

for divergence-free $\mathbf{v}$ satisfying the appropriate boundary conditions. Clearly the first term in the integrand $\sim |\nabla\mathbf{v}|^2$ is positive, but the second term $\sim v_1 v_3$ is of indefinite sign, and it is easy to see that the positive terms win out at low Reynolds numbers, corresponding to large ν or small U. To establish this we estimate the relative magnitudes of these two terms. According to the Schwarz inequality and the relation $2|ab| \leq a^2 + b^2$,

$$\left| \int_0^{L_1} dx \int_0^{L_2} dy \int_0^h dz\, v_1 v_3 \right| \leq \|v_1\|_2 \, \|v_3\|_2 \leq \frac{1}{2}\|\mathbf{v}\|_2^2. \tag{10.2.33}$$

Because each component of $\mathbf{v}$ is periodic in x and y and vanishes at $z = 0$ and $z = h$, Poincaré's inequality (Theorem 2.1) implies

$$\|\mathbf{v}\|_2^2 \leq \frac{h^2}{\pi^2}\|\nabla\mathbf{v}\|_2^2. \tag{10.2.34}$$

Then,

$$\begin{aligned} G_\phi\{\mathbf{v}\} &\geq \frac{1}{2}\nu\|\nabla\mathbf{v}\|_2^2 - \frac{Uh^2}{2h\pi^2}\|\nabla\mathbf{v}\|_2^2 \\ &= \frac{\nu}{2}\left(1 - \frac{R}{\pi^2}\right)\|\nabla\mathbf{v}\|_2^2. \end{aligned} \tag{10.2.35}$$

Thus if $R \leq \pi^2 \sim 10$, then the Couette profile satisfies the spectral condition and we can assert that the laminar dissipation rate is an upper bound as well as a lower bound. (This crude estimate of R below which the laminar flow is realized is very conservative.) When R is larger than this, then this argument no longer ensures that the spectral constraint is satisfied by Couette flow and other profiles must be produced to provide an upper bound.

The technical challenge for larger R is to choose a profile that satisfies the boundary conditions but does not violate the spectral constraint.

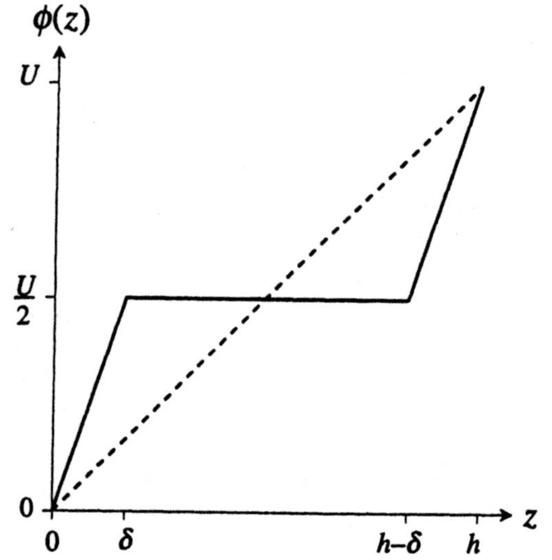

Fig. 10.1. Background flow profile $\phi(z)$. The linear Couette flow profile is shown for comparison (dashed line).

Referring to (10.2.28), it is apparent that this will be possible only if the $|\nabla \mathbf{v}|_2^2$ term dominates the $\phi' v_1 v_3$ term for the relevant class of vector fields $\mathbf{v}$. A profile with $\phi' = 0$ would suffice, but then ϕ could not satisfy its boundary conditions. The boundary conditions on $\mathbf{v}$ demand that v_1 and v_3 vanish at $z = 0$ and $z = h$. Thus it may be possible to satisfy the spectral constraint with a choice of profile where the shear rate $\phi'(z)$ is small over most of the interval $[0, h]$, but where the necessarily nonvanishing values of ϕ' are concentrated toward the ends of the interval where v_1 and v_3 are relatively small.

These observations may be implemented in the background flow profile

$$\phi(z) = \begin{cases} Uz/2\delta, & 0 \le z \le \delta \\ U/2, & \delta \le z \le h - \delta \\ \frac{U}{2\delta}(z - h + 2\delta), & h - \delta \le z \le h \end{cases} \tag{10.2.36}$$

illustrated in Figure 10.1. We refer to the parameter δ as the "boundary layer thickness." At a given value of R, δ will be adjusted to ensure that the spectral constraint will be satisfied. The smaller δ is chosen, the more positive G_ϕ will be, but at the expense of a poorer (larger) upper bound on ε.

To see how this works, estimate the $\phi' v_1 v_2$ term in G_ϕ as follows. Application of the fundamental theorem of calculus and the Schwarz inequality shows that the $x - y$ integral of the product $v_1 v_3$ is bounded uniformly in z according to

$$\left| \int_0^{L_1} dx \int_0^{L_2} dy\, v_1(x,y,z) v_3(x,y,z) \right| \leq z \sqrt{\int_0^{L_1} dx \int_0^{L_2} dy \int_0^{z} dz \left(\frac{\partial v_1}{\partial z} \right)^2} \times \sqrt{\int_0^{L_1} dx \int_0^{L_2} dy \int_0^{z} dz \left(\frac{\partial v_3}{\partial z} \right)^2}. \tag{10.2.37}$$

An analogous estimate holds near the $z = h$ boundary. The $\int \phi' v_1 v_3$ term in G_ϕ is then simply estimated in terms of δ and the $\|\nabla \mathbf{v}\|_2^2$ term:

$$\begin{aligned} &\left| \int_0^{L_1} dx \int_0^{L_2} dy \int_0^{h} dz\, \phi' v_1 v_3 \right| \\ = &\ \frac{U}{2\delta} \left| \int_0^{L_1} dx \int_0^{L_2} dy \int_0^{\delta} dz\, v_1 v_3 + \int_0^{L_1} dx \int_0^{L_2} dy \int_{h-\delta}^{h} dz\, v_1 v_3 \right| \\ \leq &\ \frac{U\delta}{4} \left\{ \frac{1}{2\sqrt{2}} \left\| \frac{\partial v_1}{\partial z} \right\|_2^2 + \frac{\sqrt{2}}{2} \left\| \frac{\partial v_3}{\partial z} \right\|_2^2 \right\} \\ \leq &\ \frac{U\delta}{8\sqrt{2}} \|\nabla \mathbf{v}\|_2^2. \end{aligned} \tag{10.2.38}$$

The incompressibility constraint on $\mathbf{v}$ has been used in the last step above (see exercise 10.1). Thus,

$$G_\phi\{\mathbf{v}\} \geq \frac{\nu}{2} \|\nabla \mathbf{v}\|_2^2 - \frac{U\delta}{8\sqrt{2}} \|\nabla \mathbf{v}\|_2^2 = \frac{\nu}{2} \left(1 - \frac{U\delta}{4\sqrt{2}\nu} \right) \|\nabla \mathbf{v}\|_2^2, \tag{10.2.39}$$

and the boundary layer thickness may be adjusted so that the spectral condition is fulfilled by choosing

$$\delta = 4\sqrt{2} \frac{\nu}{U} = 4\sqrt{2} h R^{-1}. \tag{10.2.40}$$

The corresponding upper bound on the energy dissipation rate is

$$\varepsilon \leq \frac{\nu}{h} \int_0^h |\phi'(z)|^2\, dz = \frac{1}{8\sqrt{2}} \frac{U^3}{h}. \tag{10.2.41}$$

This is a rigorous upper bound on the energy dissipation, valid when $R \geq 8\sqrt{2}$ so that $\delta \leq h/2$. It is an interesting result for a number of reasons. First, it is *independent of viscosity* in accord with Kolmogorov's scaling view of turbulent energy dissipation (recall the discussion in section 3.3). In addition, the prefactor $\left(8\sqrt{2}\right)^{-1} \approx 0.088$ is substantially less than $O(1)$, hinting that this result is more than just a dimensional analysis argument. This is pursued further in section 10.4. Moreover, our analysis, and in particular the form of the background profile in (10.2.36), hints at the generic boundary layer structure of high Reynolds number flows (recall the discussion in section 3.2). Here we are forced to consider such a flow configuration by the spectral constraint.

The upper bound on the drag force implied by (10.2.41) is

$$F = \tau \times A = \frac{\varepsilon h A}{U} \leq \frac{U^2 A}{8\sqrt{2}}. \tag{10.2.42}$$

In nondimensional terms we have

$$\frac{h^2 \tau}{\rho \nu^2} \leq \frac{U^2 h^2}{8\sqrt{2}\nu^2} = \frac{1}{8\sqrt{2}} R^2. \tag{10.2.43}$$

Summarizing the results of this section, we have proved that for the boundary-driven shear flow considered, the time averaged viscous drag is bounded from above and below in terms of the Reynolds number according to

$$R \leq \frac{h^2 \tau}{\rho \nu^2} \leq \frac{1}{8\sqrt{2}} R^2. \tag{10.2.44}$$

10.3 Thermal convection in a horizontal plane

The nondimensional formulation of the Boussinesq equations for the velocity, pressure, and temperature fields in the case of a horizontal layer of fluid heated from below is (recall section 2.1)

$$\frac{\partial \mathbf{u}}{\partial t} + \mathbf{u} \cdot \nabla \mathbf{u} + \nabla p = \sigma \Delta \mathbf{u} + \hat{\mathbf{k}} \sigma Ra T, \tag{10.3.1}$$

$$\nabla \cdot \mathbf{u} = 0 \tag{10.3.2}$$

$$\frac{\partial T}{\partial t} + \mathbf{u} \cdot \nabla T = \Delta T, \tag{10.3.3}$$

where σ is the Prandtl number for a fluid with kinematic viscosity ν and thermal conduction coefficient κ,

$$\sigma = \frac{\nu}{\kappa}, \tag{10.3.4}$$

and Ra is the Rayleigh number appropriate for a temperature drop δT across a layer of thickness h of fluid with density ρ and thermal expansion coefficient α in a gravitational field of strength g:

$$Ra = \frac{\alpha g \delta T h^3}{\rho \nu \kappa}. \tag{10.3.5}$$

The boundary conditions on $\mathbf{u}$ in (10.3.1)-(10.3.2) are no-slip at the bottom ($z = 0$) and top ($z = 1$), i.e., $\mathbf{u}(x, y, 0, t) = \mathbf{u}(x, y, 1, t) = 0$. The boundary conditions on T correspond to unit temperature difference across the gap, $T(x, y, 0, t) = 1$ and $T(x, y, 1, t) = 0$. We will impose periodic boundary conditions of dimensionless lengths λ_1 and λ_2 in the horizontal directions. This situation is illustrated in Figure 1.6.

The natural measure of the heat flow is the ratio of the total rate of heat transport to the conductive rate of heat transport (recall (1.5.17) and (2.1.36)). This ratio is the *Nusselt number* which we define here as

$$Nu = 1 + \left\langle \frac{1}{A} \int_0^{\lambda_1} dx \int_0^{\lambda_2} dy \int_0^1 dz\, u_3 T \right\rangle, \tag{10.3.6}$$

where $A = \lambda_1 \lambda_2$ is the horizontal area and $\langle \cdot \rangle$ means the largest possible time average, as in (6.3.10). The challenge is to deduce the $Nu - Ra - \sigma$ relationship,

$$Nu = Nu(Ra, \sigma) \tag{10.3.7}$$

from the equations of motion (10.3.1) and (10.3.2).

The connection between the heat transport and the energy dissipation rate is obtained from the energy evolution equation for the velocity field. Dotting $\mathbf{u}$ into (10.3.2), integrating over the volume and then integrating by parts using the divergence-free condition, we find

$$\frac{1}{2}\frac{d}{dt}\|\mathbf{u}\|_2^2 = -\sigma \|\nabla \mathbf{u}\|_2^2 + \sigma\, Ra \int_0^{\lambda_1} dx \int_0^{\lambda_2} dy \int_0^1 dz\, u_3 T. \tag{10.3.8}$$

The time averaged dissipation rate is thus

$$\varepsilon = \frac{\langle \|\nabla \mathbf{u}\|_2^2 \rangle}{A} = Ra(Nu - 1). \tag{10.3.9}$$

Hence bounds on ε in terms of Ra (and possibly σ) yield bounds on Nu in terms of Ra (and possibly σ).

For this problem the best lower bound follows immediately from (10.3.9). Because ε and Ra are manifestly nonnegative, it must be that $Nu \geq 1$. The lower bound $Nu = 1$ is realized in the pure conduction state where $\varepsilon = 0$, the temperature has a linear profile across the gap, and there is no fluid motion. This is the unique asymptotic state of the fluid at Rayleigh numbers below the critical Rayleigh number corresponding to the loss of nonlinear stability (see exercise 2.3). This lower bound may also be derived from a variational formulation, which is a useful exercise here because it will lead the way to a variational principle for an upper bound.

The deviation θ of the temperature field from the linear conduction profile, defined by

$$T(x,y,z,t) = 1 - z + \theta(x,y,z,t), \tag{10.3.10}$$

evolves according to

$$\frac{\partial \theta}{\partial t} + \mathbf{u} \cdot \nabla \theta = \Delta \theta + u_3. \tag{10.3.11}$$

This deviation vanishes on the top and bottom plates, and so satisfies the boundary conditions $\theta(x,y,0,t) = \theta(x,y,1,t) = 0$. The evolution of the L^2 norm of θ is

$$\frac{1}{2}\frac{d}{dt}\|\theta\|_2^2 = -\|\nabla\theta\|_2^2 + \int_0^{\lambda_1} dx \int_0^{\lambda_2} dy \int_0^1 dz\, u_3\theta. \tag{10.3.12}$$

The boundary and divergence-free conditions on $\mathbf{u}$ imply that the volume integral of u_3 against any function of z alone vanishes identically (see exercise 10.2), so

$$\int_0^{\lambda_1} dx \int_0^{\lambda_2} dy \int_0^1 dz\, u_3\theta = \int_0^{\lambda_1} dx \int_0^{\lambda_2} dy \int_0^1 dz\, u_3 T. \tag{10.3.13}$$

Because $\nabla\theta = \nabla T + \hat{\mathbf{k}}$,

$$\|\nabla\theta\|_2^2 = \|\nabla T\|_2^2 - A. \tag{10.3.14}$$

Hence

$$\frac{1}{2}\frac{d}{dt}\|\theta\|_2^2 = -\|\nabla T\|_2^2 + A + \int_0^{\lambda_1} dx \int_0^{\lambda_2} dy \int_0^1 dz\, u_3 T, \tag{10.3.15}$$

and if $\|\theta\|_2^2$ remains bounded uniformly in time as $t \to \infty$ (which can be shown: see exercise 10.3), then the time average of (10.3.15) yields

$$Nu = \frac{\langle \|\nabla T\|_2^2 \rangle}{A}. \tag{10.3.16}$$

That is, the Nusselt number is just the time average of the square of the L^2 norm of the gradient of the temperature field.

That the pure conduction state is the state of minimum Nu then follows from the variational principle

$$\|\nabla T\|_2^2 \geq \inf \left\{ \|\nabla\tau\|_2^2 \;\middle|\; \tau(x,y,0) = 1, \tau(x,y,1) = 0, \tau \text{ periodic in } x, y \right\}. \tag{10.3.17}$$

The Euler-Lagrange equation for the extremal field, $0 = \delta\|\nabla\tau\|_2^2/\delta\tau$ is Laplace's equation

$$\Delta\tau = 0 \tag{10.3.18}$$

with the boundary conditions given in (10.3.17). The minimizing solution is the conduction profile $\tau = 1 - z$.

A variational problem can also be posed for upper bounds on Nu. Inspired by the analysis in the previous section, we decompose the temperature field into a stationary background profile $\tau(z)$ and a fluctuation $\theta(x, y, z, t)$ as

$$T(x, y, z, t) = \tau(z) + \theta(x, y, z, t), \tag{10.3.19}$$

where the inhomogeneous boundary conditions are contained in τ. That is, $\tau(0) = 1$ and $\tau(1) = 0$, so the fluctuations vanish at the top and the bottom; $\theta(x, y, 0, t) = \theta(x, y, 1, t) = 0$.

The evolution equation for θ is

$$\frac{\partial\theta}{\partial t} + \mathbf{u}\cdot\nabla\theta = \Delta\theta + \tau'' - \tau' u_3, \tag{10.3.20}$$

where τ' and τ'' are derivatives of $\tau(z)$. The square of the L^2 norm of ∇T is

$$\|\nabla T\|_2^2 = \|\nabla\theta\|_2^2 + 2\int_0^{\lambda_1} dx \int_0^{\lambda_2} dy \int_0^1 dz\, \tau' \frac{\partial\theta}{\partial z} + A\int_0^1 dz\, [\tau'(z)]^2. \tag{10.3.21}$$

The evolution equation for the L^2 norm of the fluctuating field derived from (10.3.20) is

$$\begin{aligned} \frac{1}{2}\frac{d}{dt}\|\theta\|_2^2 &= -\|\nabla\theta\|_2^2 - \int_0^{\lambda_1} dx \int_0^{\lambda_2} dy \int_0^1 dz\, \tau' \frac{\partial\theta}{\partial z} \\ &\quad - \int_0^{\lambda_1} dx \int_0^{\lambda_2} dy \int_0^1 dz\, \tau' u_3 \theta, \end{aligned} \tag{10.3.22}$$

and using (10.3.21) in (10.3.22) we find

$$\begin{aligned} \frac{1}{2}\frac{d}{dt}\|\theta\|_2^2 + \frac{1}{2}\|\nabla T\|_2^2 &= \frac{1}{2}A\int_0^1 dz\, [\tau'(z)]^2 \\ &\quad - \int_0^{\lambda_1} dx \int_0^{\lambda_2} dy \int_0^1 dz \left(\frac{1}{2}|\nabla\theta|^2 + \tau' u_3 \theta\right). \end{aligned} \tag{10.3.23}$$

The kinetic energy evolution equation is

$$\frac{1}{2}\frac{d}{dt}\|\mathbf{u}\|_2^2 = -\sigma\|\nabla\mathbf{u}\|_2^2 + \sigma\, Ra \int_0^{\lambda_1} dx \int_0^{\lambda_2} dy \int_0^1 dz\, u_3\theta, \tag{10.3.24}$$

so combining (10.3.23) with (10.3.24) we arrive at

$$\frac{1}{2}\frac{d}{dt}\left[\|\theta\|_2^2+(\sigma Ra)^{-1}\|\mathbf{u}\|_2^2\right]+\frac{1}{2}\|\nabla T\|_2^2=\frac{1}{2}A\int_0^1 dz\,[\tau'(z)]^2$$
$$-\int_0^{\lambda_1}dx\int_0^{\lambda_2}dy\int_0^1 dz\left(\frac{1}{2}|\nabla\theta|^2+(\tau'-1)u_3\theta+Ra^{-1}|\nabla\mathbf{u}|^2\right). \quad (10.3.25)$$

This is the analogue of (10.2.22) for the shear flow problem in the previous section and we proceed from this point along similar lines.

The functional $H_\tau\{\theta,\mathbf{u}\}$ is defined by

$$H_\tau\{\theta,\mathbf{u}\}=\int_0^{\lambda_1}dx\int_0^{\lambda_2}dy\int_0^1 dz\left(\frac{1}{2}|\nabla\theta|^2+(\tau'-1)u_3\theta+Ra^{-1}|\nabla\mathbf{u}|^2\right) \quad (10.3.26)$$

for divergence-free vector fields $\mathbf{u}(x,y,z)$ and functions $\theta(x,y,z)$ which are periodic in x and y and which also vanish at the top and bottom plates. *If* this is a nonnegative functional, then we may drop the last term in (10.3.25) and change the equation into a differential inequality,

$$\frac{1}{2}\frac{d}{dt}\left[\|\theta\|_2^2+(\sigma Ra)^{-1}\|\mathbf{u}\|_2^2\right]+\frac{1}{2}\|\nabla T\|_2^2\le\frac{1}{2}A\int_0^1 dz\,[\tau'(z)]^2. \quad (10.3.27)$$

As long as initial values of $\|\theta\|_2^2$ and $\|\mathbf{u}\|_2^2$ are finite, this allows us to conclude that the largest possible time average of $\|\nabla T\|_2^2$ is bounded by $\|\nabla\tau\|_2^2$:

$$\left\langle\|\nabla T\|_2^2\right\rangle\le A\int_0^1 dz\,[\tau'(z)]^2. \quad (10.3.28)$$

We summarize these observations in the form of a variational principle:

$$Nu\le\inf\left\{\int_0^1[\tau'(z)]^2\,dz\;\middle|\;\tau(0)=1,\tau(1)=0\text{ and }H_\tau\{\theta,\mathbf{u}\}\ge 0\right\}. \quad (10.3.29)$$

This is the same kind of constrained variational problem encountered in section 10.2 for the shear driven flow example. The condition $H_\tau\ge 0$ is the "spectral constraint" for this problem and, as in the shear flow case, the extremal profile is not generally known. The utility of this formulation is its ability to provide upper bounds on Nu for trial profiles that satisfy the boundary conditions and the spectral constraint. Note that the spectral constraint only contains *Ra*, so the Prandtl number has dropped out of this formulation.

The linear conduction profile, $1-z$, satisfies the boundary conditions and is the extremum profile if it satisfies the spectral constraint. This is precisely the case for small Rayleigh numbers, the calculation being the same as the nonlinear stability analysis for this situation. This is in accord with the nonlinear stability of the pure conduction state at low *Ra*. At higher *Ra*, however, another test profile must be used.

The structure of acceptable profiles at high *Ra* can be motivated by considering the terms that contribute to H_τ. The $|\nabla\theta|^2$ and $|\nabla\mathbf{u}|^2$ terms are always positive, whereas the middle term, $(\tau'(z)-1)u_3\theta$, is indefinite. If we could choose $\tau' = 1$ all across the gap then the middle term would vanish and H_τ would be manifestly positive, but then τ's boundary conditions couldn't be satisfied because they imply that the average of τ' across the gap is -1. On the other hand, the vanishing boundary conditions on θ and $\mathbf{u}$ at the plates imply that θ and u_3 are relatively small near the boundaries. So if the background temperature profile can be chosen so that $\tau' - 1$ is concentrated near the top and bottom boundaries, then there is a chance that the integrals of the positive gradient terms will overwhelm the integral of $(\tau' - 1)u_3\theta$, rendering H_τ nonnegative.

Based on these remarks we consider trial profiles of the form illustrated in Figure 10.2,

$$\tau(z) = \begin{cases} 1 + \left(1 - \delta^{-1}\right) z, & 0 \le z \le \delta \\ z, & \delta \le z \le 1 - \delta \\ \left(1 - \delta^{-1}\right)(z - 1), & 1 - \delta \le z \le 1. \end{cases} \tag{10.3.30}$$

The "thermal boundary layer" thickness δ may be adjusted to ensure that $H_\tau \ge 0$. Indeed, estimating the terms in H_τ as in the last section, we may find that for any $\alpha > 0$,

$$H_\tau\{\theta, \mathbf{u}\} \ge \frac{1}{2}\left(1 - \frac{\delta\alpha}{2}\right)\|\nabla\theta\|_2^2 + \left(Ra^{-1} - \frac{\delta}{8\alpha}\right)\|\nabla\mathbf{u}\|_2^2. \tag{10.3.31}$$

The choices $\alpha = 2\delta^{-1}$ and

$$\delta = 4Ra^{-1/2} \tag{10.3.32}$$

ensure that $H_\tau\{\theta, \mathbf{u}\} \ge 0$. The Nusselt number is then bounded from above as

$$Nu \le \int_0^1 [\tau'(z)]^2\, dz = \frac{1}{2}Ra^{1/2} - 3, \tag{10.3.33}$$

valid in the parameter regime $Ra \ge 64$ so that $\delta \le 1/2$ and $Nu \ge 1$.

10.4 Discussion

Questions of existence, uniqueness, and regularity of solutions of the Navier-Stokes equations in the *absence* of boundaries have been explored

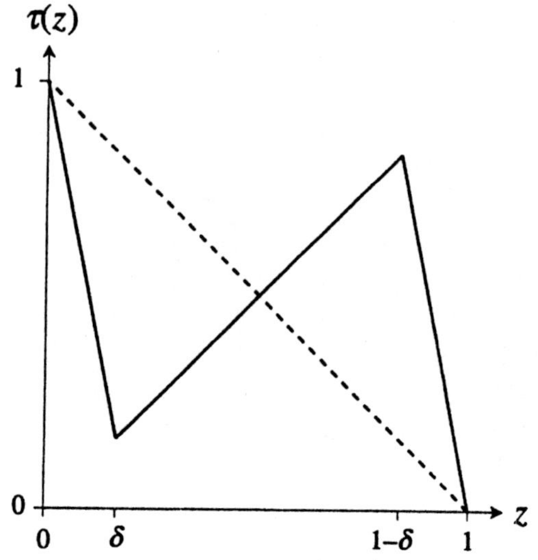

Fig. 10.2. Background temperature profile $\tau(z)$. The linear conduction profile is shown for comparison (dashed line).

in considerable depth in previous chapters. In this chapter we studied flows driven by inhomogeneous Dirichlet (no-slip) boundary conditions. For completeness, we remark here on the technical points associated with related issues of existence, uniqueness, and regularity in the presence of boundaries.

Weak solutions of the Navier-Stokes or Boussinesq equations can be constructed in many cases where some no-slip and/or inhomogeneous boundary conditions are imposed by the same general approach as discussed in Chapter 5 for periodic boundary conditions: A sequence of finite dimensional Galerkin approximations – modal truncations – is constructed and shown to possess an appropriately convergent subsequence. In the completely periodic (no boundary) case it is natural to use Fourier modes, the eigenfunctions of the Laplacian, as the basis for the Galerkin approximations. In the presence of nontrivial boundary conditions, some other modes which satisfy the appropriate boundary conditions must be employed. For the Navier-Stokes equations a natural choice of basis is the complete set of eigenfunctions of the Stokes operator, which is the product of the projector onto divergence-free vector fields and the Laplacian, acting on divergence-free vector fields. On simple periodic domains the projection operator and spatial derivatives commute and the Stokes operator is the same as the familiar Laplacian with the spectrum simply restricted appropriately to divergence-free fields. This is not the case for general boundary conditions, and the Stokes operator can have a spectrum which is not a subset of the Laplacian on scalars (see exercise 10.4).

Mathematically, the essential technical difficulty with rigid boundaries is that the projection operator onto divergent-free vector fields does not always commute with spatial derivatives. The practical manifestation of this is the appearance of boundary terms in the course of performing the integrations by parts used to derive the differential inequalities. In turn, these are normally used to establish a priori bounds on appropriate seminorms of the solutions of, say, some Galerkin approximations. These boundary terms may prevent the construction of the ladderlike structure of Chapter 6 which connects together norms of various derivatives of the fields, and so prevents us from making analagous statements about higher derivatives of the fields.

Physically, rigid boundaries produce boundary layers. These are regions in space near the no-slip (or isothermal) walls where the boundary conditions, rather than the evolution equations, determine the structure of the flow (or temperature) field. It should not be surprising that the resulting competition between the bulk dynamics and the boundary constraints can lead to additional physical complexity and corresponding analytical complications.

Weak solutions of the Navier-Stokes or Boussinesq equations can be constructed for the cases considered in sections 10.2 and 10.3. They will be square integrable functions pointwise in time, and the square of the L^2 norm of their first derivatives will be locally integrable in time. That is, a Leray-type inequality (see (5.3.41)) will be valid for the weak solutions so we can be assured that the time integrals of the energy dissipation rates are indeed finite. Although the weak solutions may not be smooth enough for the various manipulations of the equations of motion that we performed in the last two sections to be valid, the Galerkin approximations will be smooth enough, and they will satisfy appropriate energy evolution equations from which the conclusions of the preceding sections follow. The question of uniqueness of the weak solutions remains and, in general, this is unresolved. The time averaged energy dissipation rate estimates derived here, however, are valid for all weak solutions whether or not they are unique for some specific initial conditions. The point of these remarks is that we can be confident that the bounds, based on energy estimates, on energy dissipation rates deduced in sections 10.2 and 10.3 apply to weak solutions of the dynamical equations.

It is also of interest to see how the rigorous analytical bounds derived here compare with approximate theories and/or experiments. The standard statistical turbulence theory of high Reynolds number boundary-driven shear flow was developed in section 3.1. Detailed predictions for the drag as a function of the Reynolds number were given there: In Figure 10.3 we plot the rigorous upper and lower bounds derived in this chapter along with the results of the mixing length closure approximation to the Reynolds equation. The conventional theory predicts the logarithmic friction law for the drag with asymptotic behavior as $R \to \infty$,

$$\frac{h^2\tau}{\rho v^2} \to \kappa^2 \frac{R^2}{(\log R)^2} \approx 0.16 \frac{R^2}{(\log R)^2}, \tag{10.4.1}$$

where the von Karman constant κ is a fitting parameter of the theory with the nominal value of ~ 0.4. This does not violate the rigorous bound derived in section 10.2 (see equation (10.2.44)),

$$\frac{h^2\tau}{\rho v^2} \leq \frac{1}{8\sqrt{2}} R^2 \approx 0.088 R^2. \tag{10.4.2}$$

The rigorous upper bound has the same Reynolds number exponent and a comparable prefactor, but with the additional logarithmic factors.

Both the prediction of the approximate theory and these results of the rigorous analysis can be compared with recent high Reynolds number experiments on turbulent flow between concentric cylinders, at least in the limit of large aspect ratio and a narrow gap. The conventional mixing length closure approximation to the Reynolds equation, i.e., the logarithmic friction law, fits the experimental data quite well in the range $Ra \sim 10^4 - 10^6$, leading to the conjecture that it may be quantitatively capturing the asymptotic behavior. If this is the case, then the upper estimate in (10.4.2) – at least so far as the Reynolds number dependence is concerned – may be sharp to within logarithms. (This can be compared with the analysis in Chapter 9 where the dimension of the attractor for the $2d$ system is bounded from above to within logarithms of lower estimates.) Such results are encouraging, but it is natural to ask where there might be room for improvement.

Ideally the variational problem (10.3.29) for the optimal background flow profile would be solved exactly. This would yield the best estimates that this method has to offer. In this chapter, however, two short cuts have been taken in order to obtain explicit results. First, trial profiles

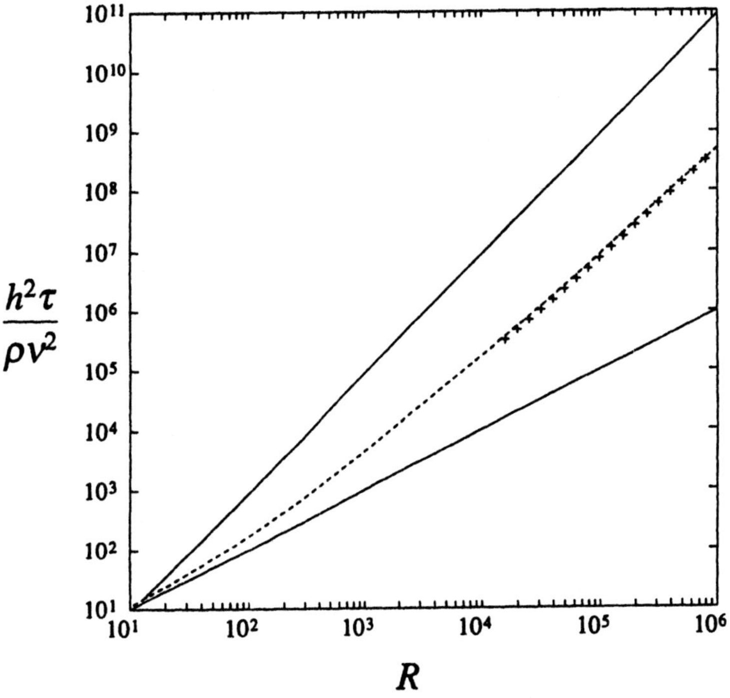

Fig. 10.3. Drag versus Reynolds number. The solid lines are the rigorous upper and lower bounds from (10.2.15). The dashed line is the result of the closure approximation in Chapter 3 ($\kappa = 0.4$) and the discrete data are a fit to experimental measurements for turbulent flow between concentric cylinders (Lathrop, Fineberg, and Swinney 1992).

were sampled only from a very restricted class of functions, namely, simple piecewise linear functions as in Figure 10.1. This was done for analytical convenience. Second, the spectral condition on the profile was not explicitly verified; rather, elementary and somewhat crude estimates were employed to ensure that the constraint was satisfied. In all likelihood the constraint is over-satisfied and we have produced an overestimate of the best bound for ε. What has been neglected in the verification of the spectral contraint for the piecewise linear profiles in section 10.2 is the divergence-free restriction on the functions in the domain of H_{ϕ}. To optimally check the spectral constraint for a given test background profile, the eigenvalue problem in (10.3.33) should really be solved exactly, and the lowest eigenvalue should be determined as a functional of ϕ. In the case of piecewise linear $\phi(z)$, the eigenvalue problem is a set of linear, piecewise constant coefficient differential equations. These equations have been solved exactly in the $2d$ situation ($u_2 = v_2 = 0; \partial/\partial y = 0$) for the piecewise linear profiles, yielding the same R^2 power law for the drag but with the prefactor reduced by more than an order of magnitude. A next step will be to consider other test profiles, for example, the approximate mean flow profiles as in (3.2.29). We stress, however, that the extremum profile of the variational problem will not be a mean flow in the usual sense of statistical turbulence theory. The real question for the shear flow problem is this: Can we deduce the functional form of the logarithmic friction law rigorously, as an upper bound, directly from the incompressible Navier-Stokes equations? The same question is relevant to channel and pipe flow problems where a similar variational approach can be developed, and where the logarithmic friction law has also been successfully applied to experiments.

It remains a challenge to develop similar background flow decompositions and variational bounds for open systems such as grid generated turbulence or for flow past a solid object. In those cases, experiments anticipate scaling of the energy dissipation rate similar to that found in section 10.2, i.e., $\varepsilon \sim U^3/L$ for appropriate velocity and length scales, but *without* the apparent logarithmic corrections that arise in wall bounded shear flows.

In the case of the convection problem, recent experiments, theory, and numerical simulations indicate crossovers from

$$Nu \sim Ra^{1/3} \tag{10.4.3}$$

to

$$Nu \sim Ra^{2/7} \tag{10.4.4}$$

at higher Rayleigh numbers, with a final asymptotic regime where $Nu \sim Ra^{1/2}$ with logarithmic corrections. The rigorous result derived in section 10.3 is

$$Nu \leq c\, Ra^{1/2}, \tag{10.4.5}$$

consistent to within $\log Ra$ factors of the expected asymptotic behavior. There is room for improvement along a number of lines for the convection problem. The points mentioned above for the shear flow problem, i.e., fully employing the divergence-free condition in the spectral constraint and utilizing a broader class of background profiles, apply to the convection setup as well. Additionally, it is currently an open question whether the inclusion of a background flow profile, along with the background temperature profile, can improve the bounds. The idea is that large scale convection roll structures are neglected in the analysis developed in section 10.3, and these may play a dominant role in the Nu–Ra relationship.

An interesting case, which to date has not been studied in the context of the background flow method, is infinite Prandtl number convection. Infinite Prandtl number convection in a horizontal plane is modeled by the dynamical equations

$$\nabla p = \Delta \mathbf{u} + \hat{\mathbf{k}} Ra T, \tag{10.4.6}$$

$$\nabla \cdot \mathbf{u} = 0, \tag{10.4.7}$$

$$\frac{\partial T}{\partial t} + \mathbf{u} \cdot \nabla T = \Delta T, \tag{10.4.8}$$

with similar boundary conditions at the top and bottom plates: $\mathbf{u}(x, y, 0, t) = \mathbf{u}(x, y, 1, t) = 0$. The velocity field is (linearly but nonlocally) slaved to the temperature field in this system. A Reynolds decomposition approach predicts a bound of $\sim Ra^{1/3}$ on Nu in this model, and it is interesting to note the intimate velocity temperature connection in these equations, anticipating the introduction of a nontrivial background flow field in the variational formulation.

In conclusion, the results presented in this chapter illustrate the potential utility of rigorous analysis of the Navier-Stokes and related equations in problems of direct physical and engineering relevance. And although many mathematical challenges remain, the outlook is hopeful for continued development of the techniques developed here, resulting in improved analytical estimates and more fundamental physical understanding of the structure and behavior of highly stressed fluid dynamical systems.

10.5 References and further reading

The results in this chapter were first derived by Doering and Constantin [50, 51, 52]. The idea of dealing with inhomogeneous boundary conditions by introducing a "background" decomposition goes back to Hopf [54]. A variational approach for the analysis of mean flow structures based on the Reynolds decomposition was developed earlier this century (see the review by Howard [55]). The predictions of those studies, in terms of both the Reynolds and Rayleigh number scaling and the magnitudes of the prefactors, are generally the same as those derived directly from the Navier-Stokes and Boussinesq equations by the methods of this chapter. Recent experiments on turbulent flow between concentric cylinders are reported in Lathrop, Fineberg, and Swinney [57] and [58]. High Rayleigh number turbulent heat conduction experiments and theories are described, respectively, in Heslot, Castaing, and Libchaber [59] and Castaing, Gunaratne, Heslot, Kadanoff, Libchaber, Thomae, Wu, Zaleski, and Zanetti [60].

Exercises

1. Show that if $\mathbf{v}$ is periodic in x and y and vanishes at $z = 0$ and $z = h$ and $\nabla \cdot \mathbf{v} = 0$, then

$$\left\|\frac{\partial v_3}{\partial z}\right\|_2^2 \leq \left\|\frac{\partial v_1}{\partial x}\right\|_2^2 + \left\|\frac{\partial v_2}{\partial y}\right\|_2^2 + \left\|\frac{\partial v_1}{\partial y}\right\|_2^2 + \left\|\frac{\partial v_2}{\partial x}\right\|_2^2. \tag{E10.1}$$

2. In the context of section 10.3, show that if $f = f(z)$ is a function of z alone, then

$$\int_0^{\lambda_1} dx \int_0^{\lambda_2} dy \int_0^1 dz\, u_3(x, y, z, t) f(z) = 0. \tag{E10.2}$$

3. Show that $\|\theta(\cdot, t)\|_2^2$ is bounded uniformly in time.
4. Find the eigenfunctions and the eigenvalues of the Stokes problem

$$\lambda_1 v_1 = -\Delta v_1 + \frac{\partial p}{\partial x} \tag{E10.3}$$

$$\lambda_1 v_2 = -\Delta v_2 + \frac{\partial p}{\partial y} \tag{E10.4}$$

$$0 = \frac{\partial v_1}{\partial x} + \frac{\partial v_2}{\partial y} \tag{E10.5}$$

for $x \in [0, L_1]$ and $y \in [0, L_2]$, with Dirichlet boundary conditions in the y-direction, with $\mathbf{v}(x, 0) = \mathbf{v}(x, L_2) = 0$, and periodic conditions in the x-direction. Compare the spectrum with that of $-\Delta$ on scalars on the same domain with the same boundary conditions.

Appendix A

Inequalities

In this appendix we collect together some of the inequalities that are used in the text. Some are very elementary and some are not so elementary. Some we prove below, some we have proved in detail in the text, and some we merely quote. For those that we don't derive explicitly, references are provided.

Triangle inequality. If x and y are real or complex numbers or vectors,

$$|x+y| \leq |x| + |y|. \tag{A.0.1}$$

Arithmetic-mean-Geometric-mean inequality. If a and b are real numbers,

$$ab \leq \frac{1}{2}(a^2+b^2). \tag{A.0.2}$$

Moreover, if $a_1, \ldots, a_N$ are real, positive numbers, then

$$\prod_{n=1}^{N} a_n \leq \frac{1}{N} \sum_{n=1}^{N} a_n^N. \tag{A.0.3}$$

Proof: Because the square of any real number is nonnegative,

$$0 \leq (a-b)^2 = a^2 + b^2 - 2ab, \tag{A.0.4}$$

and (A.0.2) follows immediately. The general result in (A.0.3) can be derived from (A.0.2) by induction on N. □

Young's inequality. If a and b are real numbers, and ε is any positive real number, then

$$ab \leq \frac{1}{2}\left(\varepsilon a^2 + \frac{b^2}{\varepsilon}\right). \tag{A.0.5}$$

Proof: Replace a with $\varepsilon^{1/2}a$ and b with $\varepsilon^{-1/2}b$ in (A.0.2). □

A version of Hölder's inequality. If a and b are real numbers, and p and q are positive real numbers satisfying $1/p + 1/q = 1$, then

$$ab \leq \frac{a^p}{p} + \frac{b^q}{q}. \tag{A.0.6}$$

Proof: We prove it for p and q rational. Let m and n be integers with $n > m$. Let $p = n/m$ and $q = n/(n-m)$. Start with the Arithmetic-mean-Geometric-mean inequality (A.0.3) for n real positive numbers, choosing the first m equal to $a^{1/m}$ and the remaining $n-m$ equal to $b^{1/(n-m)}$. Then

$$ab \leq \frac{ma^{n/m} + (n-m)b^{n/(n-m)}}{n} = \frac{a^p}{p} + \frac{b^q}{q}. \tag{A.0.7}$$

Cauchy's inequality. Let $\mathbf{x}$ and $\mathbf{y}$ be N-dimensional vectors. Then

$$\mathbf{x} \cdot \mathbf{y} \leq |\mathbf{x}|\,|\mathbf{y}|, \tag{A.0.8}$$

where the inner product is

$$\mathbf{x} \cdot \mathbf{y} = \sum_{n=1}^{N} x_n y_n, \tag{A.0.9}$$

and the norm is

$$|\mathbf{x}|^2 = \sum_{n=1}^{N} x_n^2. \tag{A.0.10}$$

Proof: Start with (A.0.2) for $a_n = x_n/|\mathbf{x}|$ and $b_n = y_n/|\mathbf{y}|$. Summing over n,

$$\frac{\mathbf{x} \cdot \mathbf{y}}{|\mathbf{x}|\,|\mathbf{y}|} = \sum_{n=1}^{N} \frac{x_n}{|\mathbf{x}|} \frac{y_n}{|\mathbf{y}|} \leq \frac{1}{2} \sum_{n=1}^{N} \left(\frac{x_n^2}{|\mathbf{x}|^2} + \frac{y_n^2}{|\mathbf{y}|^2} \right) = 1. \tag{A.0.11}$$

Cauchy-Schwarz inequality. Let f and g be square integrable functions on a domain Ω. Then

$$\int_\Omega fg\, dV \leq \left(\int_\Omega |f|^2\, dV \right)^{1/2} \left(\int_\Omega |g|^2\, dV \right)^{1/2} = \|f\|_2 \|g\|_2, \tag{A.0.12}$$

where we have introduced the L^2 norm $\|f\|_2 = \left(\int_\Omega |f|^2\, dV\right)^{1/2}$.

Hölder's inequality. Let f and g be functions on a domain Ω such that $|f|^p$ and $|g|^q$ are integrable for some positive real p and q satisfying $1/p + 1/q = 1$. Then

$$\int_\Omega fg\,dV \le \left(\int_\Omega |f|^p\,dV\right)^{1/p} \left(\int_\Omega |g|^q\,dV\right)^{1/q} = \|f\|_p \|g\|_q, \qquad \text{(A.0.13)}$$

where we have introduced the L^p and L^q norms. This is also true for $p = 1$ and $q = \infty$ where the L^∞ norm is

$$\|f\|_\infty = \text{ess-sup}_{x\in\Omega}|f(x)|, \qquad \text{(A.0.14)}$$

which, for continuous functions, is just the sup-norm.

Minkowski's inequality. Let f and g be functions on a domain Ω such that $|f|^p$ and $|g|^p$ are integrable for some positive real $p \ge 1$. Then

$$\|f + g\|_p \le \|f\|_p + \|g\|_p. \qquad \text{(A.0.15)}$$

Some calculus inequalities. Let f be a smooth, square integrable, mean zero periodic function on $\Omega = [0, L]^d$. For spatial dimensions $d = 1, 2,$ and 3, there exist finite absolute constants c_d (i.e., no dependence on L or the choice of f) such that

$$d = 1 \qquad \|f\|_\infty^2 \le c_1 \,\|\nabla f\|_2 \|f\|_2, \qquad \text{(A.0.16)}$$

$$d = 2 \qquad \|f\|_\infty^2 \le c_2 \,\|\nabla f\|_2^2 \left[1 + \log\left(\frac{L\|\Delta f\|_2}{\|\nabla f\|_2}\right)\right], \qquad \text{(A.0.17)}$$

$$d = 3 \qquad \|f\|_\infty^2 \le c_3 \,\|\Delta f\|_2 \|\nabla f\|_2. \qquad \text{(A.0.18)}$$

Proof: Inequalities (A.0.17) and (A.0.18) are proved in Chapters 7 and 8, respectively. Inequality (A.0.16) mimics those proofs, and is left as an exercise for the reader. Similar results can be proved in higher dimensions, involving norms of successively higher derivatives. □

Some more calculus inequalities (Gagliardo-Nirenberg). Let f be a smooth, square integrable, mean zero periodic function on $\Omega = [0, L]^d$. Then for $1 \le q, r \le \infty$ and j and m integers satisfying $0 \le j < m$,

$$\|D^j f\|_p \le c\,\|D^m f\|_r^a \,\|f\|_q^{1-a}, \qquad \text{(A.0.19)}$$

where

$$\frac{1}{p} = \frac{j}{d} + a\left(\frac{1}{r} - \frac{m}{d}\right) + \frac{(1-a)}{q}, \qquad \text{(A.0.20)}$$

for a in the interval

$$\frac{j}{m} \leq a < 1, \tag{A.0.21}$$

with the exception that if $m - j - d/r$ is a nonnegative integer, then a is restricted to j/m.

Proof: See the references.

Lieb-Thirring inequalities for orthogonal functions. Consider the set of orthonormal functions $\{\phi_1, \phi_2, \ldots, \phi_N\}$ in d spatial dimensions. These must obviously satisfy

$$\int_\Omega \sum_{n=1}^{N} |\phi_n|^2 \, d^d x = N. \tag{A.0.22}$$

Lieb and Thirring have shown that they obey the inequality

$$\int_\Omega \left(\sum_{n=1}^{N} |\boldsymbol{\phi}_n|^2 \right)^{\frac{d+2}{d}} d^d x \leq c \sum_{n=1}^{N} \int_\Omega |\nabla \boldsymbol{\phi}_n|^2 \, d^d x, \tag{A.0.23}$$

where the constant c is independent of N. This is also true for vector functions $\boldsymbol{\phi}_i$.

References and further reading

The book by Beckenbach and Bellman [61] gives an elementary account of the basic inequalities. Inequality (A.0.17) was first proved by Brezis and Gallouet [62]. The Gagliardo-Nirenberg calculus inequalities are proved for functions on a variety of domains in Nirenberg [63] and Adams [64]. The Lieb-Thirring inequalities are proved in [65].

References

[1] D. J. Tritton. *Physical Fluid Dynamics.* Oxford Science Publications, Clarendon Press, Oxford, second edition, 1988.

[2] C. Bardos, F. Golse, and C. D. Levermore. Fluid dynamic limits of kinetic equations. *Journal of Statistical Physics*, 63:323–344, 1991.

[3] P. G. Drazin and W. H. Reid. *Hydrodynamic Stability.* Cambridge University Press, Cambridge, 1981.

[4] B. Straughan. *The Energy Method, Stability and Nonlinear Convection.* Springer Series in Applied Mathematics, vol. 91, Berlin, 1992.

[5] E. S. Titi. On a criterion for locating stable stationary solutions of the Navier-Stokes equations. *Nonlinear Analysis: Methods and Applications,* 11:1085–1102, 1987.

[6] E. S. Titi. Un critere pour l'appoximation des solutions périodique des équations des Navier-Stokes. *C. R. Acad. Sci. Paris, Serie I*, 312:41–43, 1991.

[7] H. Tennekes and J. L. Lumley. *A First Course in Turbulence.* MIT Press, Cambridge, Mass., 1972.

[8] R. H. Kraichnan and D. Montgomery. Two-dimensional turbulence. *Reports in Progress in Physics*, 43:547–619, 1980.

[9] P. G. Drazin. *Nonlinear Systems.* Cambridge Texts in Applied Mathematics, Cambridge, 1992.

[10] J. D. Farmer, E. Ott, and J. Yorke. The dimension of chaotic attractors. *Physica*, 7D:153–180, 1983.

[11] P. Constantin and C. Foias. Global Lyapunov exponents, Kaplan-Yorke formulas and the dimension of the attractors for 2*d* Navier-Stokes equations. *Communications in Pure and Applied Mathematics*, 38:1–27, 1985.

[12] P. Constantin and C. Foias. *Navier-Stokes Equations.* The University of Chicago Press, 1988.

[13] A. V. Babin and M. I. Vishik. Attractors of partial differential evolution and estimates of their dimension. *Russian Mathematical Surveys*, 38:133–187, 1983.

[14] R. Temam. *Infinite Dimensional Dynamical Systems in Mechanics and Physics*, vol. 68 of *Applied Mathematical Sciences.* Springer-Verlag, New York, 1988.

[15] E. N. Lorenz. Deterministic nonperiodic flow. *Journal of Atmospheric Science*, 20:130–141, 1963.

[16] C. Sparrow. *The Lorenz Equations: Bifurcations, Chaos and Strange*

Attractors. Springer Series in Applied Mathematics 41. Springer, Berlin, 1982.

[17] Y. Trève. Boxing in the Lorenz attractor. Unpublished, 1979.

[18] P. Constantin, C. Foias, O. P. Manley, and R. Temam. Determining modes and fractal dimension of turbulent flows. *J. Fluid Mech.*, 150:427–440, 1985.

[19] A. Eden, C. Foias, and R. Temam. Local and global Lyapunov exponents. *J. of Dynamics and Differential Equations*, 3:133–177, 1991.

[20] W. Rudin. *Principles of Mathematical Analysis*. McGraw-Hill, 1976.

[21] W. Rudin. *Functional Analysis*. McGraw-Hill, New York, 1973.

[22] M. Reed and B. Simon. *Methods of Modern Mathematical Analysis I: Functional Analysis*. Academic Press, New York, 1980.

[23] J. Leray. Essai sur le mouvement d'un liquide visquex emplissant l'espace. *Acta Math.*, 63:193–248, 1934.

[24] R. Temam. *The Navier-Stokes Equations and Non-linear Functional Analysis*. CBMS-NSF Regional Conference Series in Applied Mathematics. SIAM, 1983.

[25] F. John. *Partial Differential Equations*. Springer, Berlin, 1982.

[26] M. Bartuccelli, C. R. Doering, J. D. Gibbon, and S. J. A. Malham. Length scales in solutions of the Navier-Stokes equations. *Nonlinearity*, 6:549–568, 1993.

[27] M. Bartuccelli, C. Doering, and J. D. Gibbon. Ladder theorems for the 2d and 3d Navier-Stokes equations on a finite periodic domain. *Nonlinearity*, 4:531–542, 1991.

[28] W. D. Henshaw, H. O. Kreiss, and L. G. Reyna. On the smallest scale for the incompressible Navier-Stokes equations. *Theoret. Comput. Fluid Dynamics*, 1:65–95, 1989.

[29] P. G. Saffman. *Vorticity*. Cambridge University Press, Cambridge, 1993.

[30] O. A. Ladyzhenskaya. *The Mathematical Theory of Viscous Incompressible Flow*. Gordon and Breach, New York, second edition, 1963.

[31] J. Serrin. *The Initial Value Problem for the Navier-Stokes Equations*. Nonlinear Problems. University of Wisconsin Press, Madison, R. E. Langer edition, 1963.

[32] J. L. Lions and E. Magenes. *Nonhomogeneous Boundary Value Problems and Applications*. Springer-Verlag, New York, 1972.

[33] J. Hale. *Asymptotic behaviour of dissipative systems*. Mathematical Surveys and Monographs Vol 25, AMS, Providence, Rhode Island, 1988.

[34] M. I. Vishik. *Asymptotic behaviour of solutions of Evolutionary Equations*. Cambridge University Press, Cambridge, 1992.

[35] C. Foias, C. Guillopé, and R. Temam. New a priori estimates for Navier-Stokes equations in dimension 3. *Comm. in Partial Diff. Equat.*, 6:329–359, 1981.

[36] J. D. Gibbon. A voyage around the Navier-Stokes equations. *Physica*, 92D:133–139, 1996.

[37] J. T. Beale, T. Kato, and A. Majda. Remarks on the breakdown of smooth solutions for the 3-D Euler equations. *Commun. Math. Phys.*, 94:61–66, 1984.

[38] J. Glimm and A. Jaffe. *Quantum Physics*. Springer, New York, first edition, 1984.

[39] A. Majda. Vorticity and the mathematical theory of incompressible fluid flow. *Comm. Pure Appl. Math.*, 39:187–220, 1986.

[40] A. Majda. Vorticity, turbulence and acoustics in fluid flow. *SIAM Review*, 33:349–388, 1991.

[41] C. Foias and R. Temam. Gevrey class regularity for the solutions of the Navier-Stokes equations. *J. Funct. Anal.*, 87:359–369, 1989.

[42] C. R. Doering and E. S. Titi. Exponential decay rate of the power spectrum for solutions of the Navier Stokes equations. *Phys. Fluids*, 7:1384-1390, 1995.

[43] P. Constantin, C. Foias, and R. Temam. Attractors representing turbulent flows. *Memoirs of AMS*, 53(314), 1985.

[44] P. Constantin, C. Foias, and R. Temam. On the dimension of the attractors in two-dimensional turbulence. *Physica D*, 30:284–296, 1988.

[45] P. Constantin. Collective L^{∞} estimates for families of functions with orthonormal derivatives. *Indiana University Math. Journal*, 36:603–615, 1987.

[46] C. R. Doering and J. D. Gibbon. A note on the Constantin-Foias-Temam attractor dimension estimate for two-dimensional turbulence. *Physica D*, 48:471–480, 1991.

[47] J. D. Gibbon and E. S. Titi. Attractor dimension and small length scale estimates for the 3*d* Navier-Stokes equations. *Nonlinearity*, 10:109-119, 1997.

[48] D. A. Jones and E. S. Titi. Upper bounds for the number of determining modes, nodes and volume elements for the Navier-Stokes equations. *Indiana University Mathematics Journal*, 42:875–887, 1993.

[49] A. A. Ilyin. Navier-Stokes equations on the rotating sphere. A simple proof of the attractor dimension estimate. *Nonlinearity*, 7:31–40, 1994.

[50] C. R. Doering and P. Constantin. Energy dissipation in shear driven turbulence. *Phys. Rev. Lett.*, 69:1648–1651, 1992.

[51] C. R. Doering and P. Constantin. Variational bounds on energy dissipation in incompressible flows I: Shear flow. *Phys. Rev. E*, 49:4087–4099, 1994.

[52] P. Constantin and C. R. Doering. Variational bounds on energy dissipation in incompressible flows II: Channel flow. *Phys. Rev. E*, 51:3192–3198, 1995.

[53] C. R. Doering and P. Constantin. Variational bounds on energy dissipation in incompressible flows III: Convection. *Phys. Rev. E*, 53:5957–5981, 1996.

[54] E. Hopf. Ein allgemeiner endliche-keitssatz der hydrodynamik. *Mathematische Annalen*, 117:764–775, 1941.

[55] L. N. Howard. Bounds on flow quantities. *Annual Reviews of Fluid Mechanics*, 4:473–494, 1972.

[56] F. H. Busse. The optimum theory of turbulence. *Advances in Applied Mechanics*, 18:77–121, 1978.

[57] D. Lathrop, J. Fineberg, and H. Swinney. Turbulent flow between concentric rotating cylinders at large Reynolds number. *Phys. Rev. Lett.*, 68:1515–1518, 1992.

[58] D. Lathrop, J. Fineberg, and H. Swinney. Transition to shear-driven turbulence in Couette-Taylor flow. *Phys. Rev.*, A46:6390–6405, 1992.

[59] F. Heslot, B. Castaing, and A. Libchaber. Transitions to turbulence in helium gas. *Phys. Rev.*, A36:5870–5873, 1987.

[60] B. Castaing, G. Gunaratne, F. Heslot, L. Kadanoff, A. Libchaber, S. Thomae, X-Z Wu, S. Zaleski, and G. Zanetti. Scaling of hard thermal turbulence in Rayleigh-Bénard convection. *Journal of Fluid Mechanics*, 204:1–30, 1989.

[61] E. Beckenbach and R. Bellman. *An Introduction to Inequalities.* Yale University Press, New Haven, 1961.

[62] H. Brezis and T. Gallouet. Nonlinear Schrödinger evolution equations. *Nonlinear Analysis, Theory, Methods and Applications*, 4(4):677–681, 1980.

[63] L. Nirenberg. On elliptic partial differential equations. *Annali della Scuola Norm. Sup.*, 13:115–162, 1959.

[64] R. A. Adams. *Sobolev Spaces.* Academic Press, New York, 1975.

[65] E. Lieb and W. Thirring. *Inequalities for the moments of eigenvalues of the Schrodinger equations and their relation to Sobolev inequalities.* Princeton University Press, Princeton New Jersey, 1976.

Index

Printed in the United Kingdom
by Lightning Source UK Ltd.
WO000000